Sustainable Civil Infrastructures

Sustainable Infrastructure impacts our well-being and day-to-day lives. The infrastructures we are building today will shape our lives tomorrow. The complex and diverse nature of the impacts due to weather extremes on transportation and civil infrastructures can be seen in our roadways, bridges, and buildings. Extreme summer temperatures, droughts, flash floods, and rising numbers of freeze-thaw cycles pose challenges for civil infrastructure and can endanger public safety. We constantly hear how civil infrastructures need constant attention, preservation, and upgrading. Such improvements and developments would obviously benefit from our desired book series that provide sustainable engineering materials and designs. The economic impact is huge and much research has been conducted worldwide. The future holds many opportunities, not only for researchers in a given country, but also for the worldwide field engineers who apply and implement these technologies. We believe that no approach can succeed if it does not unite the efforts of various engineering disciplines from all over the world under one umbrella to offer a beacon of modern solutions to the global infrastructure. Experts from the various engineering disciplines around the globe will participate in this series, including: Geotechnical, Geological, Geoscience, Petroleum, Structural, Transportation, Bridge, Infrastructure, Energy, Architectural, Chemical and Materials, and other related Engineering disciplines.

More information about this series at http://www.springer.com/series/15140

Janusz Wasowski · Tom Dijkstra
Editors

Recent Research on Engineering Geology and Geological Engineering

Proceedings of the 2nd GeoMEast International Congress and Exhibition on Sustainable Civil Infrastructures, Egypt 2018 – The Official International Congress of the Soil-Structure Interaction Group in Egypt (SSIGE)

Springer

Editors
Janusz Wasowski
Institute for Geo-Hydrological Protection IRPI
National Research Council (CNR)
Bari, Italy

Tom Dijkstra
School of Architecture, Building and Civil Engineering
Loughborough University
Loughborough, UK

ISSN 2366-3405 ISSN 2366-3413 (electronic)
Sustainable Civil Infrastructures
ISBN 978-3-030-02031-6 ISBN 978-3-030-02032-3 (eBook)
https://doi.org/10.1007/978-3-030-02032-3

Library of Congress Control Number: 2018957283

This Springer imprint is published by the registered company Springer Nature Switzerland AG
The registered company address is: Gewerbestrasse 11, 6330 Cham, Switzerland

Contents

About the Editors

Dr. Janusz Wasowski

Dr. Janusz Wasowski is a research geologist at CNR-IRPI (National Research Council-Institute for Geo-hydrological Protection) in Bari, Italy. He is also the editor in chief of Engineering Geology. Since 2011, he has held the positions of visiting professor at the Research School of Arid Environment and Climate Change, Lanzhou University, Gansu Province, China, and of science officer of the Natural Hazards Group Programme, European Geosciences Union (EGU).

He is an internationally recognised scientist in the field of engineering geology, natural hazards, and applied remote sensing. For over 25 years, his work has covered a broad spectrum of research topics ranging from slope instability and landslide assessment, collateral seismic hazards, geotechnical field investigation, and in situ monitoring, to exploitation of air-borne/space-borne remote sensing and geophysical surveying in engineering geology. He has also served as a consultant for the National Department for Civil Protection, Italy, the Government of Gansu Province, China, and the Centre National de l'Information Géo-Spatiale, Haiti, focusing on landslides and other geohazards and on the application of satellite multi-temporal interferometry for monitoring terrain deformations and infrastructure instability.

Since 2007, he has been a member of the Editorial Board of *Engineering Geology* (Elsevier) and the *Quarterly Journal of Engineering Geology and Hydrogeology* (The Geological Society, London). He is the author/co-author of over 100 articles/chapters and the guest editor of 11 special issues published in international scientific journals.

Dr. Tom Dijkstra

Tom is an engineering geologist/geomorphologist with an interest in geohazards, specifically landslides and climate change. He graduated from Utrecht University (NL) in 1987 with a degree in Physical Geography with Geomorphology and Quaternary Geology as specialisations and Soil Science and Palynology as additional subjects. In 1988, he moved from Utrecht University to the University of Leicester to join a large European-Chinese research team addressing loess landslide geohazards in the Lanzhou region of Gansu Province, China. In 2000, he was awarded his PhD at Utrecht University for his work on 'Loess slope instability in the Lanzhou region, China,' and he maintains a keen interest in geohazards research in Central China. A major aspect of his research is on climate change effects on UK slopes, including engineered earthwork transport infrastructure slopes. Research topics included developing an architecture for transport infrastructure earthworks resilience in a context of climate change, integrating a water balance model approach into transport network performance, and research into engineered earthworks deterioration. A key aspect of his research focuses on the interactions between geohazards and communities, and this has given him the opportunity to engage with multi-disciplinary research in China, India, Nepal, and the Caribbean.

Tom is on the editorial boards of Engineering Geology and the Quarterly Journal of Engineering Geology and Hydrogeology. He has 100+ publications and has been the editor of special issues on topics including geohazards and landslide research in China. He holds honorary research positions at Lanzhou University (Gansu Province, China) and the BGS.

Slope Stability Analysis and Suggestive Measures for an Active Landslide in Indian Himalaya

Shantanu Sarkar(✉), Manojit Samanta, Mahesh Sharma, and Ajay Dwivedi

Geotechnical Division, CSIR-CBRI, Roorkee, Uttrakhand, India
shantanu_cbri@yahoo.co.in

Abstract. Landslides are frequently occurring phenomena in Indian Himalayas causing considerable loss of lives and property every year. There are several active landslides along the major highways which pose recurring problems to the traffics particularly during the monsoon season. One such landslide has been studied in detail for planning and design of suitable control measures. The paper presents the geo-investigation and slope stability analysis carried out to arrive at suitable control measures. Geotechnical investigation and slope stability analysis were carried out to assess the present stability condition of the landslide. The results of the analysis have shown that the landslide is presently active and it needs suitable control measures to minimize the landslide activities. Soil nailing was one of the options found to be suitable for reinforcement of loose debris material on the uphill slope. It was observed that the stability of the slope significantly improves when soil nailing measure was considered.

1 Introduction

Landslide disasters causing loss of life and property are enormous in Indian Himalayas. The complex geology and tectonic set up supplemented by heavy rainfall and anthropogenic activities made the Himalayan region very prone to landslides. There are many locations where human lives are in danger due to landslides. A large number of slope failures have affected the major highways in the Himalayas, which are also important routes for pilgrims. One such National Highway (NH-58) Rishikesh-Badrinath road in the Alaknanda river valley is very prone to landslides particularly in the vicinity of Main Central Thrust (MCT) zone. A few hazard zones were identified as most potential zones, where number of landslides is affecting the traffic and thereby posing threats to lives (Sarkar et al. 2005). Such landslides were mapped on the high resolution CARTOSAT I remote sensing image and a small segment of this region is shown in the Fig. 1. In the present study, one active landslide in this region near Pipalkoti along the same highway has been investigated for stability analysis and suggestive control measure (Fig. 2).

J. Wasowski and T. Dijkstra (Eds.): GeoMEast 2018, SUCI, pp. 1–9, 2019.
https://doi.org/10.1007/978-3-030-02032-3_1

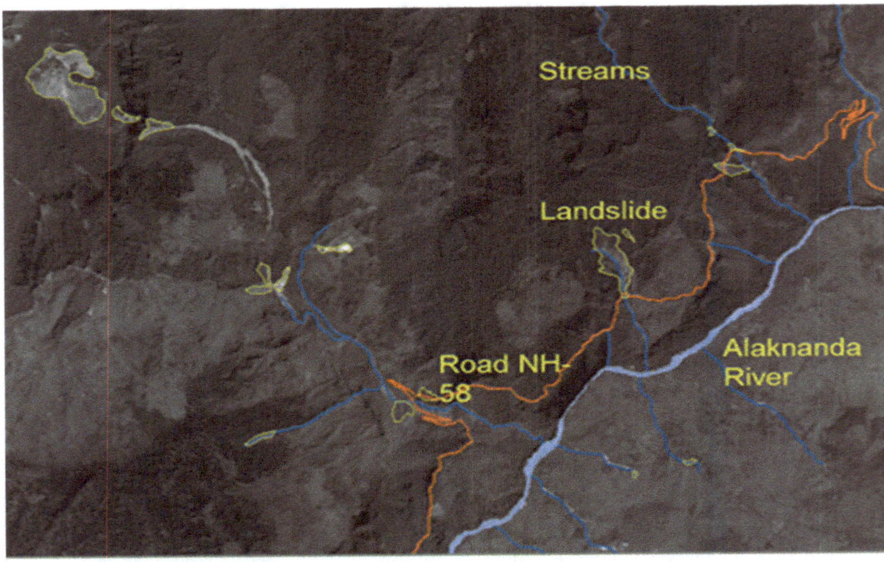

Fig. 1. Landslides along National Highway (NH-58) mapped on satellite image (Sarkar et al. 2013)

Fig. 2. A panoramic view of the landslide

2 Landslide Description

The studied landslide is a debris slide including debris materials and a few rock boulders on a steep slope. The rock types exposed in the slide area are dolomites which primarily consist of calcites and quartz. Topographic survey of the landslide was carried out to know the topographic features and to generate a contour map on 1:500 scale with 1 m contour interval. A DEM was prepared in GIS from the contour map and a slope map was generated. From the slope map it was observed that maximum area in the landslide site is having the slope angle 35–45°, however, there are a few escarpments with very steep slope of 45–60°. The landslide is divided into two parts by a seasonal drain. The road has been severely damaged on the downhill slide due to the ongoing landslide activities (Fig. 3).

Fig. 3. Highway at risk due to debris slide

The detailed sub-surface geological investigation of the landslide has indicated four different types of materials with the same litho-units of dolomitic limestone. The top layer is a soil layer having silty clay. The next layer is a debris material comprising of soil with rock fragments of dolomitic limestone which is followed by highly weathered soft dolomitic limestone. The bottom stratum is the fresh hard dolomite. The thickness of debris material was found to be around 8–10 m.

3 Geotechnical Investigation

To assess the geotechnical characteristic of the landslide site, samples were collected to evaluate its physical and engineering properties. Various test such as grain size analysis, moisture content, Atterberg limits, specify gravity, bulk density, dry density, void ratio, relative density and direct shear tests were carried out (Table 1). The grain size analysis shows that the material is dominated by gravels and sand size particle. There is no clay content which indicates that the slide material has no cohesion. Direct shear tests were carried out using a large shear box apparatus, as the slide materials contain a significant amount of gravels and sands. The direct shear tests were conducted on samples at four different normal stress (σ_n) levels. The samples were prepared in the shear box having size 300 mm × 300 mm at the minimum densities of the respective samples. The tests were conducted at normal stresses of 0.5, 1.0, 1.5 and 2.0 kg/cm^2 which approximately correspond to anticipated normal stress in the field. For a given value of the applied constant normal stress, shear stress was applied by shearing the specimen at a rate of 0.2 mm/min. The tests were conducted up to failure. The direct shear test results thus obtained are plotted (Fig. 4) to obtain shear parameters i.e. cohesion and angle of internal friction (c & Φ) which are presented in Table 2.

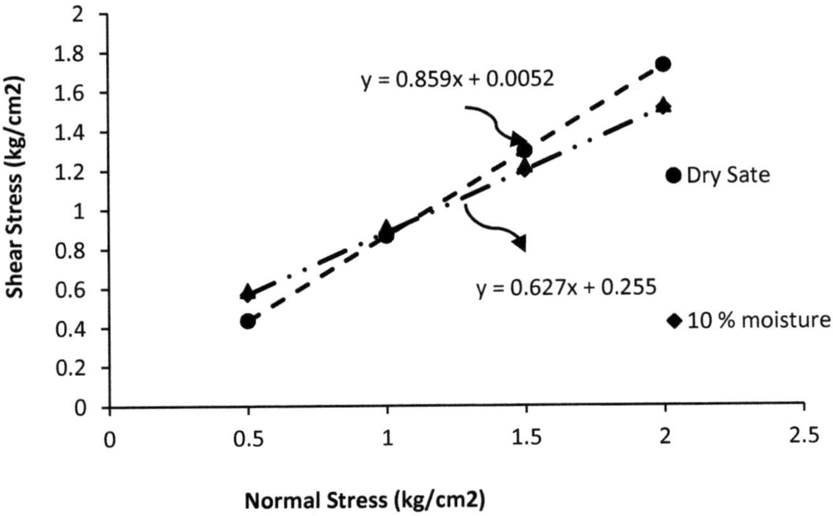

Fig. 4. Shear envelope of sample obtained from direct shear test

Table 1. Physical properties of soil samples

Sample	IS classification	Particle size distribution (%)			Physical properties		
		Gravel	Sand	Silt & Clay	Max. density	Min. density	Specific gravity
1	GP-GM	56	27	17	2.198	1.95	2.66
2	GM	42	37	21	2.04	1.81	2.67

Table 2. Results of shear test

Sample	At dry state		At 10% moisture	
	c	Φ	c	Φ
1	0.005	40.69	0.255	32.12
2	0.09	42.61	0.28	32.04

From the above data it can be said that there is an increase in cohesion 'c' and decrease in angle of internal friction "Φ" with increase in moisture content. The shear strength of the sample decreases with increase in moisture content. Hence, it can be inferred that the debris material loses its shear strength as the moisture content increase towards saturation.

4 Slope Stability Analysis and Mitigation Measures

Stability analysis of the landslide was carried out to know the state of stability condition of the landslide so that suitable control measures can be designed. The failure mechanism of the slide appeared to be a circular failure as observed from the field observations. Limit equilibrium methods were used to determine the factor of safety of the landslide slope under different condition. The analysis was carried out for two most probable failure profiles. Two sections as shown in the Fig. 5 of the slide area were selected to carry out slope stability analysis. Factor of safety were calculated using GeoStudio software (SLOPE/W) at different moisture conditions and without earthquake and with earthquake loading. The results of the analysis are shown in the Fig. 6 and the values of factor of safety are tabulated in the Table 3. From the table it can be inferred that the factors of safety of both the sections are marginally stable under dry and static conditions but it drastically decreases in the range of 0.8 to 0.5 under partial saturation and earthquake loading. Selection of an appropriate remedial measure depends on several factors such as engineering feasibility, economic feasibility, social

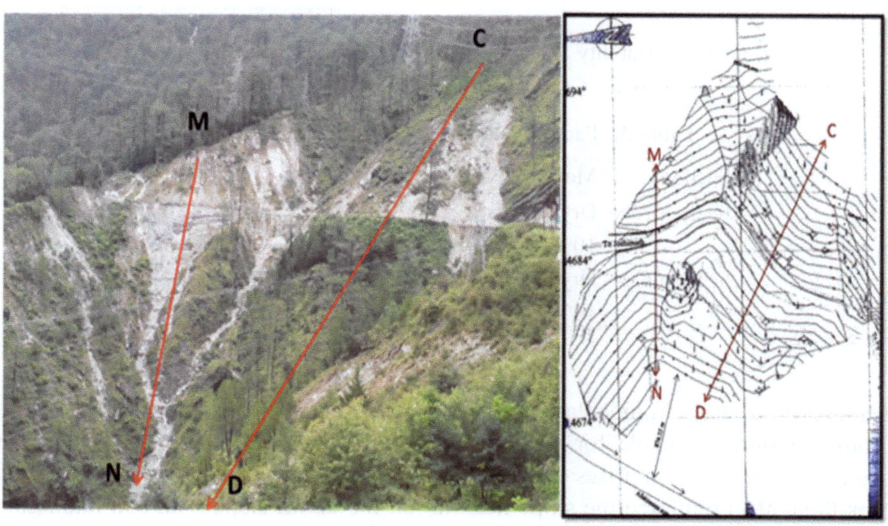

Fig. 5. Sections for stability analysis

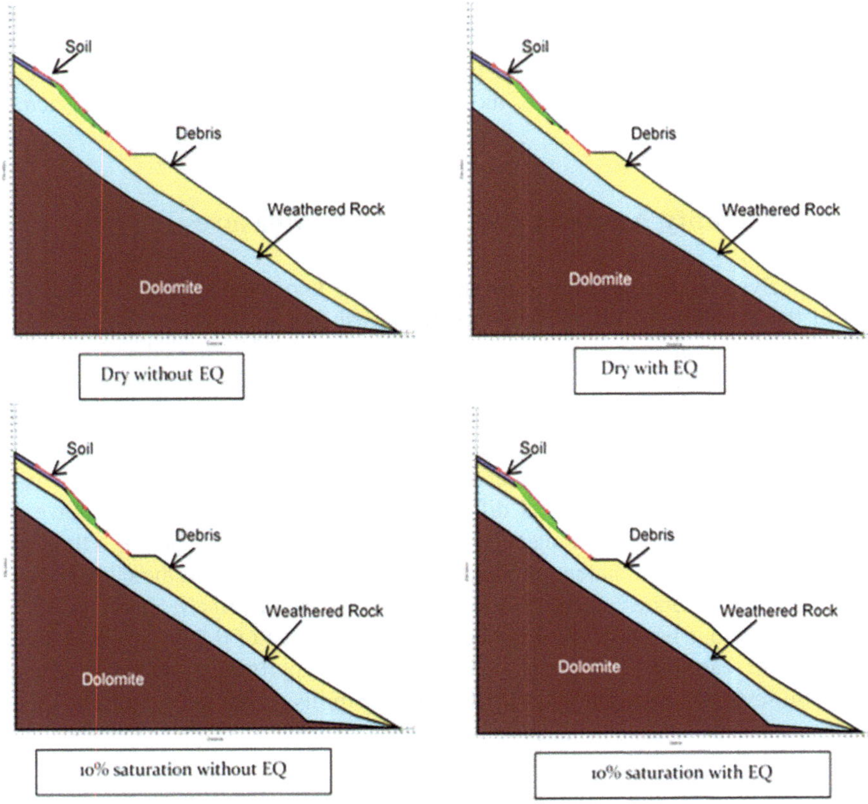

Fig. 6. Slope stability analysis without considering remedial measures

Table 3. Factor of safety of MN and CD sections

Sections	Moisture condition	Without EQ	With EQ
Section MN	Dry	0.966	0.779
	10% saturation	0.664	0.541
Section CD	Dry	1.051	0.839
	10% saturation	0.786	0.628

acceptability, and environmental acceptability (Holtz and Schuster 1996). There are various measures available for stabilizing an unstable slope. As already stated, the slope above the road comprises of loose shallow overburden of soil and debris. Hence it was thought to reinforce the slope material on the uphill slope with soil nailing.

Soil nailing is an in situ soil reinforcing technique adopted for stabilizing unstable slopes (Sharma et al. 2019). It increases the overall shear strength of unsupported soils and is relatively flexible and can accommodate large settlements. Also soil nailing is more economical than conventional concrete gravity wall. The soil nail gains pull-out resistance from within the sliding soil mass in front of the slip plane and the stable soil mass located behind the slip plane. The geometric system of soil nail placements creates an internally reinforced soil mass that is stable. The soil nail acts as a passive bearing element, which relies on soil movement and subsequent active earth pressure to mobilize the shear strength along the nail whereas a tieback anchor is pre-stressed to mobilize shear strength (Junaideen et al. 2004). Soil nails were designed suitably and stability analyses were carried out in SLOPE/W module limit equilibrium method using GeoStudio software to see the efficacy of the nails. Different parameters were analyzed by the Morgenstern-Price method. The reasonable design parameters obtained by the analysis are given in the Table 4. The values of factor of safety obtained after inclusion of soil nails are given in the Table 5 and the results of the analysis are shown in the Fig. 7. It is observed from the analysis that factor of safety increases significantly after installing soil nails as stabilization measures. The factor of safety with soil nails increase to 2.4 and 2.8 in dry condition from 0.97 and 1.0 for sections MN and CD respectively. Even under partial saturation the slope is stable after inclusion of nails.

Table 4. Specification of soil nails

Nail parameters	Values
Nail diameter	25 mm
Bar capacity	415
Bond diameter	100 mm
Spacing (horizontal & vertical)	2 m each
Nail length	5 m and 6 m
Skin friction	300 kPa

Table 5. Factor of safety of MN and CD sections with soil nailing as stabilizing measures

Sections	Moisture condition	Without EQ	With EQ
Section MN	Dry	2.388	2.210
	10% saturation	1.536	1.276
Section CD	Dry	1.723	1.410
	10% saturation	1.667	1.256

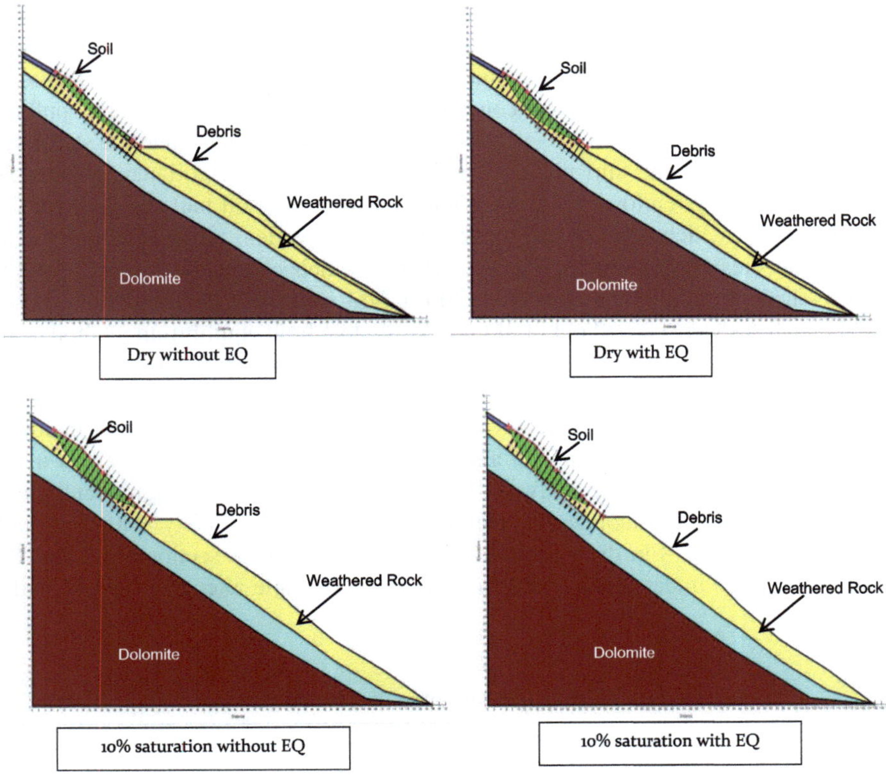

Fig. 7. Slope stability analysis with inclusion of soil nails (Geo-Slope International Ltd., GeoStudio 2007)

5 Conclusions

An active landslide along a major highway of Indian Himalayan region was investigated in detail for arriving at suitable remedial measures. Laboratory studies were conducted to determine the engineering propertied of the slope material. Stability analysis was carried out for two vulnerable sections of the landslide slope to determine the degree of instability of the slope. The factors of safety were found to be close to one which indicated the marginal stability condition of the slope. To increase the stability of the slope particularly on the uphill slope, soil nailing measure was selected. Further analysis showed that the stability of the slope has significantly increased after reinforcing the slope material with soil nails. This indicated that soil nailing could be one of the options to minimize the on-going sliding activities of the landslide. The study is being continuing for testing other measures at different locations so that a comprehensive design of control measures can be suggested.

Acknowledgments. Authors are grateful to the Director, CSIR-Central Building Research Institute for granting permission to publish the work. The assistance provided by Mr. B.S. Bisht, Mr. Saurabh Singh and Ms. Nisha is greatly acknowledged.

References

Geo-Slope International Ltd., GeoStudio: Stability modeling with SLOPE/W (2007)

Holtz, R.D., Schuster, R.L.: Stabilization of soil slopes. In: Turner, A.K., Schuster, R.L. (eds.) Landslides Investigation and Mitigation, Special Report 247, Transportation Research Board (1996)

Junaideen, S.M., et al.: Laboratory study of soil nail interaction in loose, completely decomposed granite. Can. Geotech. J. **41**(2), 274–286 (2004). https://doi.org/10.1139/t03-094

Sarkar, S., Kanungo, D.P., Patra, A.K.: Landslides in the Alaknanda valley of Garhwal Himalaya, India. Q. J. Eng. Geol. Hydrol. **39**, 79–82 (2005). https://doi.org/10.1144/1470-9236/05-020

Sarkar, S., Kanungo, D.P., Sharma, S.: Landslide hazard assessment in the upper Alaknanda valley of Indian Himalayas. Geomat.S, Nat. Hazards Risk **6**(4), 308–325 (2013). https://doi.org/10.1080/19475705.2013.847501

Sharma, M., Samanta, M., Sarkar, S.: Soil nailing: an effective slope stabilization technique. In: Pradhan, S., Vishal, V., Singh, T. (eds.) Landslides: Theory, Practice and Modelling. Advances in Natural and Technological Hazards Research, vol. 50. Springer, Cham (2019)

Analysis of a Combined Circular–Toppling Slope Failure in an Open–Pit

Maged Al Mandalawi[1(✉)], Greg You[1], Peter Dahlhaus[1], Kim Dowling[1], and Mohannad Sabry[2]

[1] Faculty of Science and Technology, Federation University, Ballarat, VIC, Australia
sp.group@ymail.com

[2] School of Civil Engineering, Western Sydney University, Kingswood, Australia

Abstract. Most studies of rock slope failures on open–pit mines have considered either toppling or circular failure stability analysis. By comparison, complex circular–toppling failure has received much less attention in the published literature. This paper presents a study using a range of methods to investigate a failure that occurred in July 2008 in Handlebar Hill, an open–pit base metal mine, near Mt Isa, Australia. Circular failure is the typical slope failure mechanism in slopes with low–strength rocks, although direct/flexural toppling of jointed columns can also occur. The study reviews circular–toppling failure mechanisms in the context of the local geotechnical and geo–hydrological conditions, which include the interaction between fault contacts and the existing deformed rocks. General limit equilibrium methods are used to evaluate the sensitivity of slope models to rock strength parameters and the trigger mechanisms. Finite element methods are used to assess the failure mechanisms and slope displacement, and a kinematic approach is used to evaluate structurally controlled slope instability mechanisms. The results demonstrate that the most credible failure mechanism was shearing along a circular path through the upper weaker rocks (leached Magazine Shale) that in turn initiated secondary block toppling, and the progressive nature of the slope failure mechanism. The use of conventional and numerical techniques for back–analysis of the combined circular–toppling failure provided key insights into the failure mechanisms and factors controlling slope instability.

Keywords: Circular failure · Toppling failure · Factor of safety
Back–analysis · Australia

1 Introduction

Rock slope failure has long been recognised as a constraint in open–pit mining (e.g. Hoek and Bray 1981; Read and Stacey 2009) and usually involves a number of failure mechanisms including, circular, planar, wedge, toppling and combination of these. Circular failures have been perhaps the most studied, through the seminal work by Terzaghi (1943) and Skempton (1948) which assumed the location and the path of the critical surface from the material strength parameters. Their work was later modified by

J. Wasowski and T. Dijkstra (Eds.): GeoMEast 2018, SUCI, pp. 10–30, 2019.
https://doi.org/10.1007/978-3-030-02032-3_2

Janbu (1954), Bishop (1955), Morgenstern and Price (1965), Spencer (1967), Priest and Brown (1983) and others who studied the stability of circular slip surfaces using the general limit equilibrium conditions.

Similarly, toppling failure received more attention in later decades with the development of deeper cut slopes. This type of rock failure usually occurs in slopes with closely spaced geological structures that dip sub–vertical to the slope surface. This failure mode results in overturning of the impacted blocks, such as jointed columns, around a fixed base pivoting point. The pioneering work by Goodman and Bray (1976) and Evans (1981) proposed a number of models for different types of potential toppling in rock slopes, with four types of toppling modes observed in practice and laboratory tests, specifically: block, flexural, block–flexural and secondary toppling. The significant factor triggering instability is the weight of the rock, its centre of gravity and the transfer of this load into the next underlying structures in primary toppling mechanism, while in the secondary toppling mechanism of failure, external factors other than the column weight is the main factor. In later work, Adhikary et al. (1997) divided the post failure of a flexible slope model slope into three zones: fully developed plastic deformation zone with visible cracks, partially developed plastic deformation zone and elastic deformation zone.

The vast majority of studies of rock slope failures in open–pit mines have considered either toppling, planar, wedge or circular failure stability analysis (Wyllie and Mah 2004; Read and Stacey 2009). By comparison, complex circular–toppling failure has received much less attention in the published literature (Mohtarami et al. 2014) and published cases are relatively rare. Mohtarami et al. (2014) proposed a theoretical solution for combined circular–toppling failure and developed modelling software, testing their model on a case study in Iran. Wyllie and Mah (2004) mention a case study of pit–crest toppling, which resulted from a circular failure in the upper part of the slope. Alejano et al. (2010) undertook back analysis of a failure in an open–pit in Valencia, Spain, concluding that it was a combined toppling–circular failure. Their analysis concluded that the toppling and circular failure had occurred simultaneously in different part of the slope, a situation never previously reported.

This paper adds to that scant literature by presenting a back–analysis of a localised combined failure of bench slope involving both, circular failure in the upper slope and block toppling in the lower slope. The failure occurred in July 2008 on the west wall of the Handlebar Hill open–pit located in Mt. Isa, in Queensland, Australia (Fig. 1). Large scale slope failure of inter–ramp or overall slopes in the open–pit did not occur very often and slope failures occur in many instances at local scale, involving one bench or parts of other benches. Consequently, bench–scale failure deserves more attention. The study investigates the slope failure through back–analysis, using a combination of simulations in the context of the geological domains. The general limit equilibrium method was used to assess the effect of shear strength parameters, geometry and groundwater on the slope stability, and kinematic analysis of the discontinuities identifies the location and condition of potential toppling. Finite element analysis analyses were conducted to simulate the slope displacement.

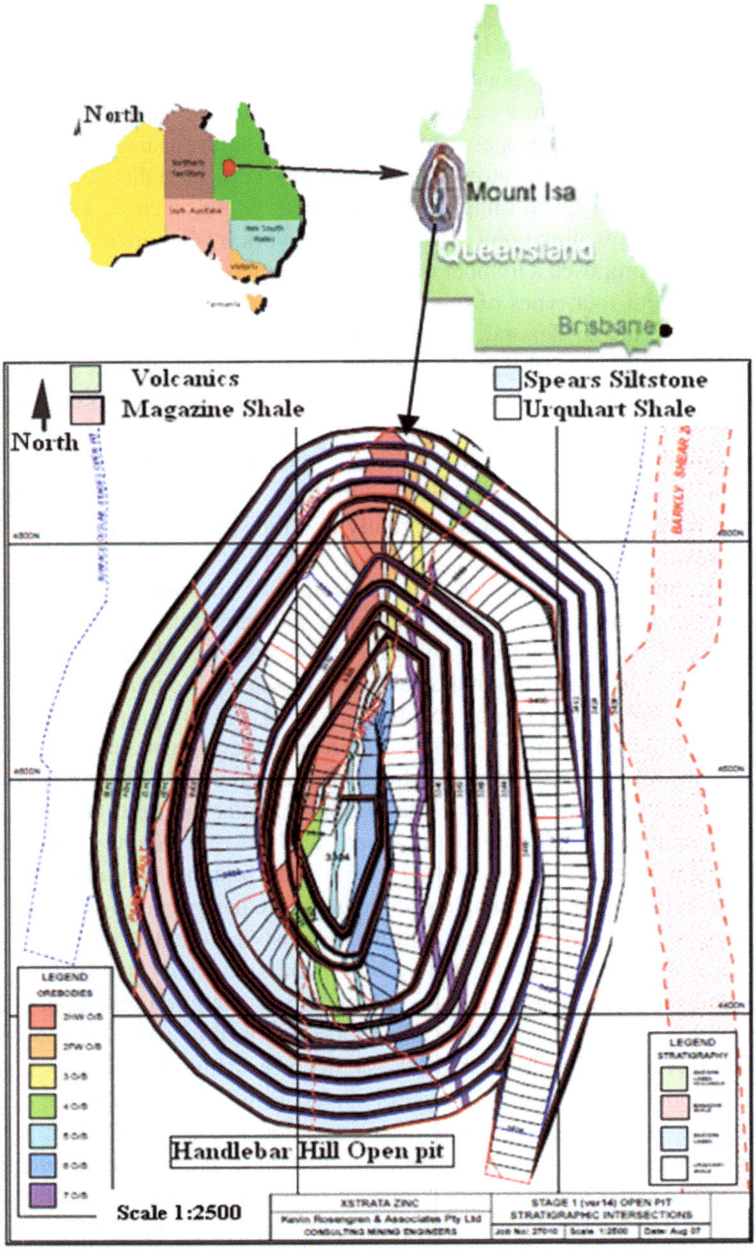

Fig. 1. Location and lithology of the first stage of excavation at the Handlebar Hill open cut mine (after Rosengren and Associates 2007)

2 Site Conditions and Engineering Geology

The Handlebar Hill open cut mine is located 20 km north of Mount Isa, an area that experiences a tropical continental climate characterised by warm drier winters and hot humid summers. Rainfall occurs mostly in the summer wet season, between November and March, however heavy rainfall can occur in any time of the year (Fig. 2). The orebody of zinc, silver and lead is within the Urquhart Shale, where seven mineralised ore zones have been identified (Rosengren and Associates 2007). A west–east cross–section of the pit (Fig. 3) shows the base of the oxidised zone, which is about 30 m to 50 m below the original ground surface, below which is the leached (altered) zone that extends to a depth of 400 m below ground surface.

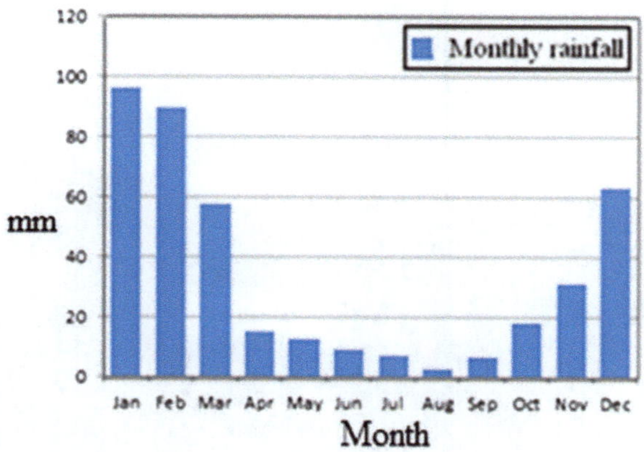

Fig. 2. Average monthly rainfalls at the site (after Rosengren and Associates 2007)

The Paroo Fault and Offset Fault are major geological structures in the west slope that occur close to the rock slope failure, known as Failure B, which is the subject of this study. Both structures dip 60°–70° to the west. The Paroo Fault is a discrete soft structure 2 m to 3 m wide with extreme shearing and clay gouge infilling. This fault intersects the west slope and acts as a contact zone between the Magazine Shale and the Volcanics. The generalised geological materials exposed in the west slope of the pit are shown in Fig. 4. At the site of the slope failure, the Magazine Shale occurs in outcrop in a relatively narrow zone near the pit slope surface and is up to 60 m thick, overlying the fresh Spears Siltstone which is up to 65 m thick. The Magazine Shale is uncomfortably overlain by fresh and weathered Volcanics, which blanket the entire west slope of the open–pit.

From the area around the failure, about fifty minor geological structures have been identified. Within the Magazine Shale zone, joint sets are the most common discontinuity, followed by offset faults, bedding and shear planes. The highly fractured nature of the mechanically weathered Magazine Shale (Fig. 5) tends to behave similarly to soil and would feasibly fail in a circular mode. Based on experimental test results from rock samples, the mechanical behaviour of the Magazine Shale was modelled using

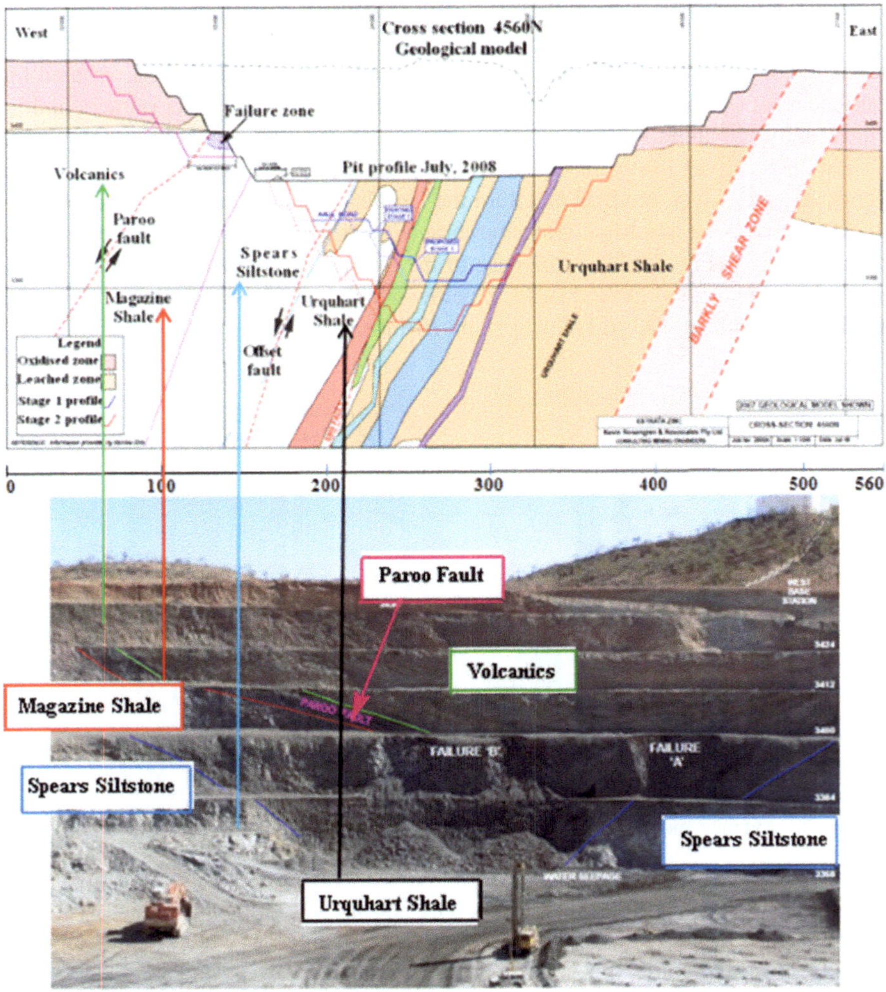

Fig. 3. East–west cross section of the open-pit correlated with a photograph of the zone of Failure B located at the west slope. Note that this cross-section is not orthogonal to the west slope (after Rosengren and Associates 2008)

RocLab 1.0 (Rocscience 2014). This model (Fig. 6) shows the relationship between normal stress and shear strength of the maximum, average, minimum (the red, black and blue curves) and weathered (the green curve) samples. This unconfined compressive strength (UCS) characterisation of the materials was defined using the generalised Hoek–Brown failure criterion. Unsurprisingly, the shear strength was the lowest in the weathered rock sample compared to other samples.

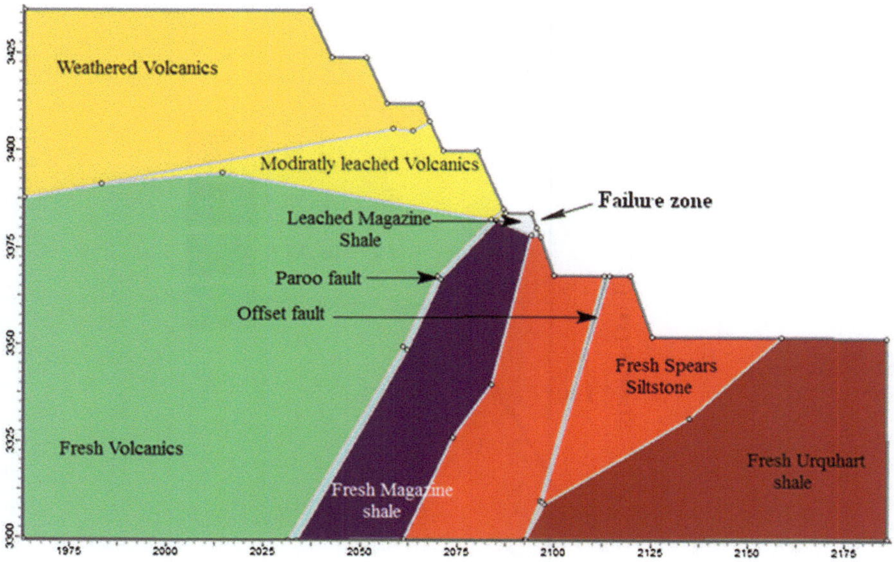

Fig. 4. General schematic of the domains of the west slope of the Handlebar Hill open–pit during the first stage of excavation and the location of the failure zone

Fig. 5. Highly fractured nature of the Magazine Shale

3 Description of the Failure

The failure occurred in June 2008 at a time when the Mt. Isa district was experiencing dry conditions, and no rainfall was recorded in June. The leached zone in the rock slope is regarded as an unconfined aquifer and some water was visible suggesting that groundwater pressures may have played some role in the failure. Weeks after the failure, water was seeping from below the northern part of the failure debris on the pit floor (after Rosengren and Associates 2007).

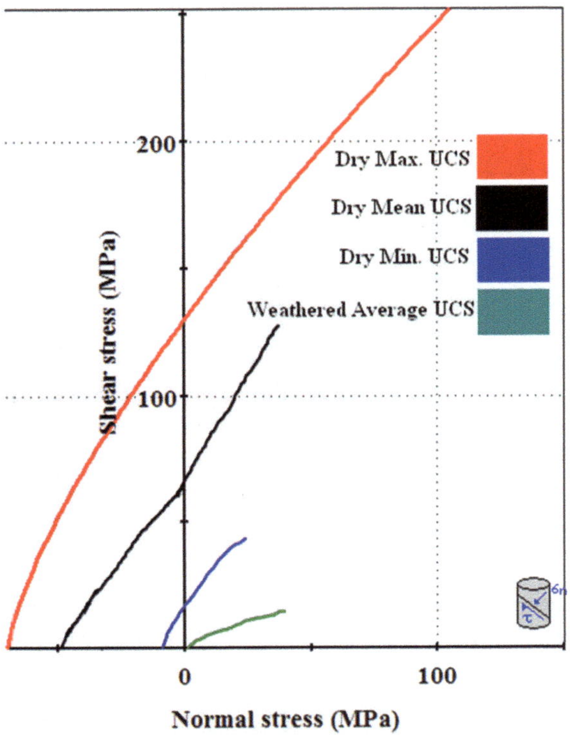

Fig. 6. Comparison of the shear strength of the Magazine Shale plotted as shear–normal stresses using Hoek–Brown strength criterion

The instability occurred when the pit depth was 80 m below the original ground surface and the pit wall dipped at up to 40° (Rosengren and Associates 2007). Before the failure event, the mining operation was still in progress to extract the uppermost part of the south orebody. The failure zone was restricted to within the Magazine Shale domain and adjacent to the upper contact with the Paroo Fault. The failure was constrained within the benches known as RL3,400 and RL3,384, but failed material spilled over the RL3,384 berm on to the pit floor at RL3,368. A slope stability monitoring prism, installed to detect the rock movement, was buried by the fallen debris (Fig. 7).

The circular failure path initiated in the upper of the Magazine Shale, where the major clay coated shear surface is a splay from the Paroo Fault. There is evidence of opening of the Magazine Shale–Paroo Fault contact behind the main failure zone (Fig. 8). The circular failure occurs by rotational failure through soil and weak, weathered rock which includes randomly oriented discontinuities, in which the failure can pass through intact rock and structural weakness (Warren and Watters 2008).

Block toppling was also recognised below the Paroo Fault, involving jointed columns that forced the lower blocks to slide (Rosengren and Associates 2008). The mass of material moved in the circular failure appears to have loaded the strata dipping into the slope, enhancing the toppling movement of the lower blocks. Therefore, the entire

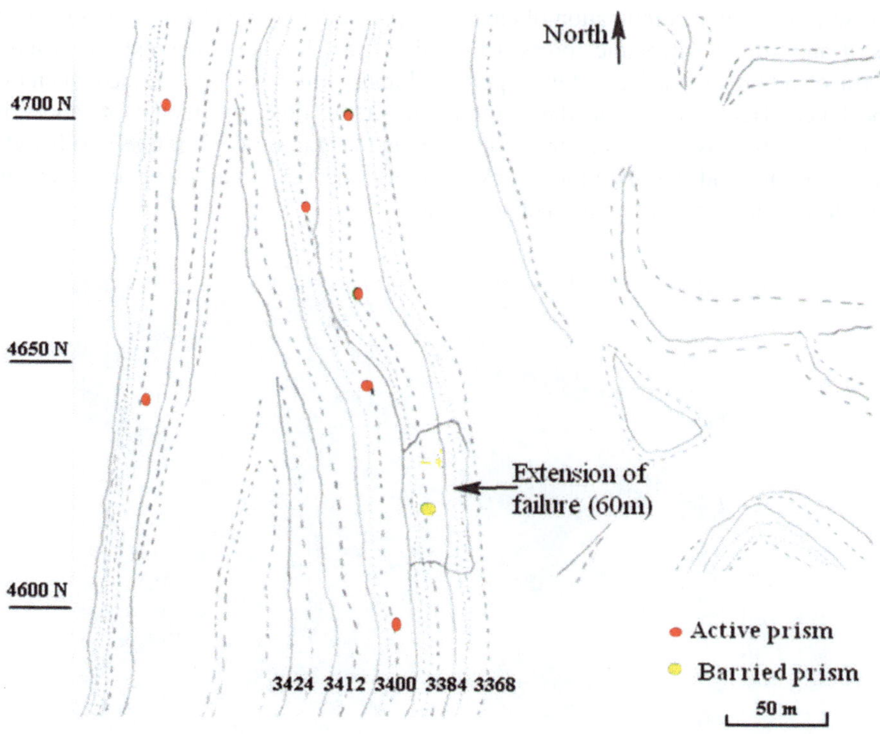

Fig. 7. Area of failure impact and measurement of rock slope displacement used for this study shows the extension of failure, bench levels involved and zone of spilling over debris

Fig. 8. Movement along contact between Magazine Shale and Paroo Fault zone above RL3,400, looking south (after Rosengren and Associates 2008)

slope went through a combination of circular–toppling failure mechanism, as shown in Figs. 9 and 10. The dynamic forces acting along the discontinuities appear to have deformed and broken the base of the toppled columns, which were overturned when the slope failed. Hence, based on the observations at the time, the model of failure is divided into two distinct mechanisms: (1) the circular failure causing the weathered–broken rock to load the uppermost rock columns, and (2) the toppling forces acting between the rock columns in an inclined plane.

Fig. 9. Top of failure on RL3,400, July 17, 2008, looking north (after Rosengren and Associates 2008)

A Slope Stability Radar (SSR) was installed in the open–pit to monitor the west slope and the monitoring system recorded movement in the prism located on the RL3,384 berm of the west slope. This prism sent a warning signal, recorded as a rock movement of 0.13 m, 36 h before the failure. This movement represents the maximum displacement recorded by the monitoring system, however it was not clear how this movement occurred, because site at the RL3,384 appears to be intact rock.

Pre–mining groundwater level data indicated that groundwater levels coincided with the base of the oxidized zone at a depth of about 50 m below ground level. This suggested that water depressurisation of the slope was not required and the water table level was not elevated by rainfall. Although it is believed that the slope was only partly saturated before the failure occurred (Rosengren and Associates 2007), there is no available evidence to confirm the groundwater pressures at the time.

Fig. 10. Close up view of rear boundary shear, looking south (after Rosengren and Associates 2008)

4 Back Analyses: Methods and Results

Given a known (or assumed) failure surface, back analysis can be carried out in order to determine the material shear strength, pore pressure or other conditions at the time of failure. The back analyses properties can be used to design remedial slope stability measures and insure best stability into the future.

For the back analysis of the circular failure, the simplified Bishop method (1955) was adapted. This method provides the scope to include an appropriate factor of safety of the slope and sensitivity to changes in mechanical properties and water pressure conditions present at the site. Furthermore, this method can be used in the combined failure of circular and toppling mechanisms, where part of the slide surface is at a soft rock interface and the shape of the slide surface differs significantly from a simple circular–slip to toppling blocks.

In circular/rotational failure models, slopes are usually comprised of low strength materials and the critical failure surface is free to take the line of least resistance through the slope (Hoek and Bray 1981). Wyllie and Mah (2004) have described the conditions for circular failure as requiring weak rock without geological structure, or highly weathered rock mass without significant structures. This description is consistent with the quality of the rock at the Handlebar Hill pit that initiated the movement at the upper part of the failure zone, that is: sheared, highly fractured and leached. Since circular failure was originally recognised in the upper RL3,400 bench and block toppling in the lower bench at RL3,384 (Rosengren and Associates 2007), this coalesces two modes of failures into a complex combined circular–toppling rock failure.

4.1 Probabilistic Back–Analysis of Material Strength

The shear strength of rock materials are related to cohesion (c′) and friction angle (φ'), and the values of these two parameters control the location of critical surface. Duncan (1996) stated that in homogenous slopes with material exhibiting an φ' greater than zero, the critical surface usually passes through the toe of the slope. Keeping the value of φ' constant and increasing the value of c′ will result in changing the location of the critical failure surface to extend deeper into the floor and away from the lower most toe. Conversely, increasing the value of φ' will reduce the radius of the critical surface within the slope and therefore, its position shrinks closer to the slope crest. So, the location of critical failure surface is controlled by the combination of φ' and c′ values of the slope rock mass.

Since the location of failure surface was located at the boundary between the Magazine Shale and Paroo Fault, which is typically weathered (leached) Magazine Shale, the values of both φ' and c′ listed in Table 1 were used within probabilistic analysis to define a relationship and estimate a minimum slope safety factor of 1 for the failure surface.

To qualify the results of the limit equilibrium method, a sensitivity analysis was carried out using a Monte Carlo probabilistic method and the results are plotted in Fig. 11. The φ' and c′ represent the strength parameters along the failure surface for a Bishop's safety factor of 1.0 for the leached Magazine Shale. Comparing these values with the parameters of rock domains (Table 1), it is clear that shear resistance within this layer was low and hence has potential for development a circular failure surface. The sensitivity plot illustrates that c′ of 64 kPa and φ' of 15° could result in the bench failure. The shear resistance properties of Magazine Shale are expected to become higher towards the innermost of the bench slope, since rock properties can vary at bench scale, and benches are more sensitive to weathering and water–runoff, which weakens these parameters (Hantz et al. 1988).

4.2 Back–Analysis of Groundwater Conditions

Although the failure mechanism is relatively complex the triggering mechanism may have been due to changes in water pressure behind the failure zone, and hence this should be considered. Limit equilibrium slope stability analysis can test the scenario that elevated pore water pressure in the slope played a part in trigging the initial failure. For this, the non–circular path search method was used in the slope model of Slide 6.0 (Rocscience 2014). A number of 10,000 possible critical surfaces were analysed within the path search option for each section using Mohr–Coulomb failure criterion. From this, the most critical surface with a minimum safety factor in the leached and the fresh Magazine Shale rock were determined. These analyses were carried out using the conventional Bishop simplified method of dividing the entire slope into a number of vertical slices and calculating the applied forces on each, using the parameters listed in Table 1.

For the entire slope, a factor of safety (FOS) of 2.197 was obtained, which adequately represents the stability condition of the overall slope prior the failure (Fig. 12). A tension crack parameter was then added to the model consistent with the NNE striking discontinuity shown in Figs. 8 and 9 and the observations of water flow after the failure.

The piezometric water levels were adjusted within the slope until the closest model of the slope failure mechanism was obtained at FOS of 0.998 (Fig. 13). However it is unlikely that this model represents the actual failure conditions at the time, because it

Table 1. Input parameters for back-analysis study

Parameters	Unit	Domain						
		Weathered Volcanics	Moderately Oxidised Volcanics	Fresh Volcanics	Leached Magazine Shale	Fresh Magazine Shale	Fresh Spears Siltstone	Fresh Urquhart Shale
γ	KN/m^3	27.4	27.6	28.3	26.5	26.5	27.2	31.1
E_{rm}	MPa	2873.13	2880	9931.5	28.94	4676	8425.46	11806.7
φ	(°)	26	28	34.6	25	28.3	37.6	37.6
c	MPa	0.284	0.295	0.714	0.100	0.339	0.896	1.088
σ_t	MPa	0.058	0.066	0.344	0.00056	0.097	0.438	0.564
σ_{ci}	MPa	36	36	55	3.0	31.8	110	108
υ	–	0.3	0.25	0.2	0.25	0.2	0.2	0.2
m_b	–	0.184	0.184	0.42	0.49	0.271	0.319	1.036
m_i	–	4	4	4.07	4	4	4	9.6
s	–	0.0003	0.0003	0.0026	0.0009	0.0008	0.0013	0.002
a	–	0.509	0.509	0.503	0.544	0.505	0.504	0.504
D	–	0.7	0.7	0.7	0.7	0.7	0.7	0.7
GSI	–	44	44	59	20	51	54	57
Dilation	(°)	5	5	8	4	5	12	12

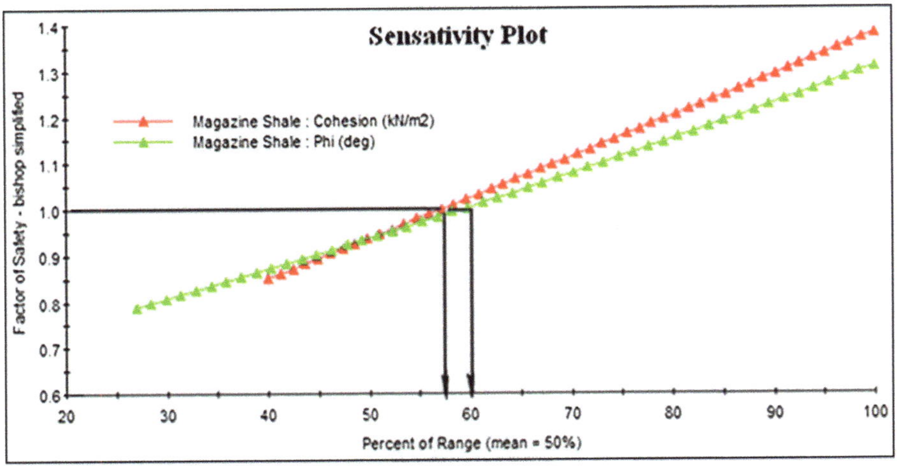

Fig. 11. Sensitivity plot of the cohesion and friction angle of the weathered Magazine Shale layer, derived from a Monte Carlo probabilistic method

does not correspond to the groundwater levels observed in previous hydrological studies. The model was adjusted to use a piezometric level of 22 m below the original ground surface, as seen during the same time of the year in earlier studies, resulting in an FOS of 2.197 (Fig. 13).

4.3 Toppling Limit Equilibrium Back-Analysis

Toppling analysis was performed using the general limit equilibrium block toppling method of Goodman and Bray (1976). This method was applied at RL3,400, which is generally weathered (leached) Magazine Shale, where the slope is 16 m high, with a bench face angle of 70° and an overall slope angle of 43°. The discontinuities are mainly joints dipping into the slope at 63° as a mean value, with a joint spacing around ≤0.25 m and lengths in the range of 2 m to 5 m for minor and major joints respectively.

In the initial model, the dip angle of the joint into the failure surface (α) requires a value of 20° to result in a minimum FOS of 1.0 (Fig. 14). In this case, the dip angle of the toppled joints into the slope is 66°. For the cases that used α = 24° and α = 26°, higher FOS values of 1.1 and 1.16 respectively resulted from the same method of analysis.

When the joints dip angle was reduced to 62° into the slope (representative of the actual conditions) and α =16°, the bench slope yielded on 5 joints under the application of Goodman and Bray (1976) method of analysis. From the top of the bench slope, the first joint where the contact circular–toppling failure occurred remained stable (Fig. 15). The second joint was unstable and experiences toppling which persisted downwards to the next four disturbed slabs. Therefore, the pressure of these toppled

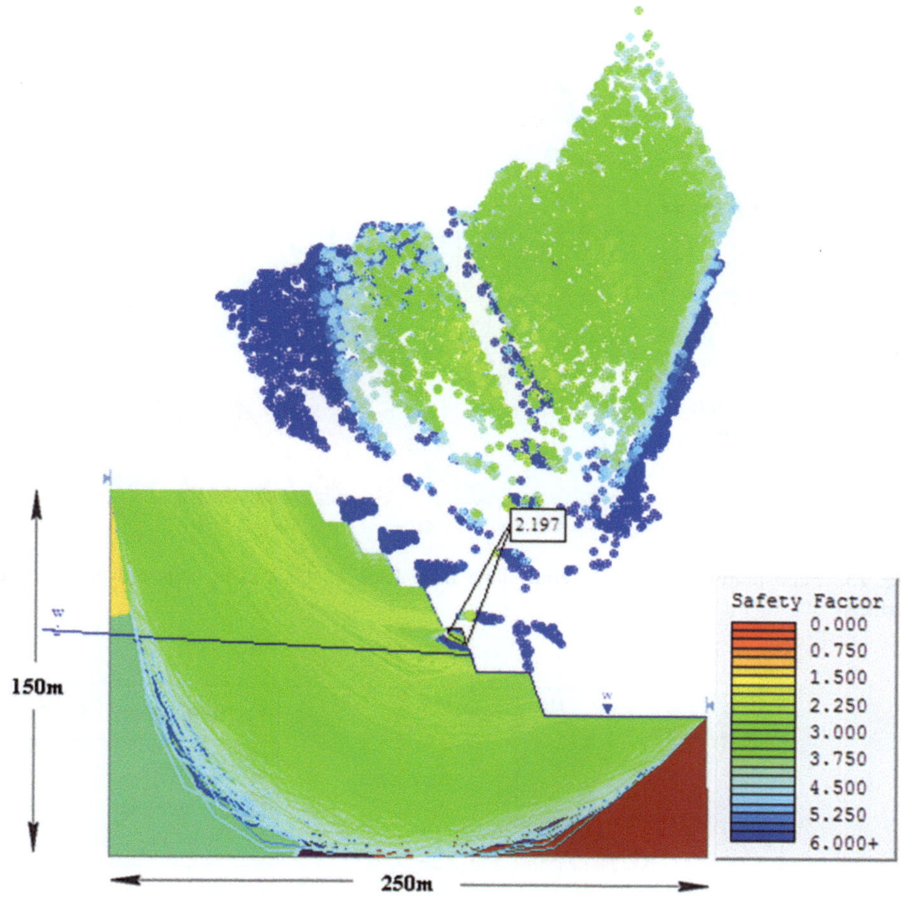

Fig. 12. Typical slope profile shows the factor of safety of 2.197 of the slope within the major zone of instability

blocks will push the last joint at the bottom of the slope to slide. The predicted toppling mechanism concurs with the observations at the time of failure, with an FOS of 0.682 (Fig. 15).

4.4 Back–Analysis Using Finite Element Method (FEM)

Back–analysis was also undertaken using two–dimensional finite element stress analysis using the Phase^2 V 9.0 (Rocscience 2014), utilising the shear strength reduction method (SSR), as an appropriate tool to investigate slope behavior with enhanced parametric flexibility. The finite element models (FEM) were conducted for both dry and partly saturated slopes utilising the same piezometric surfaces as in the limit equilibrium model. The slope was disturbed at bench RL3,400 as shown by the predicted the horizontal displacements using the generalised Hoek–Brown failure criterion for dry and partly saturated slopes respectively (Figs. 16 and 17). Generally, increasing

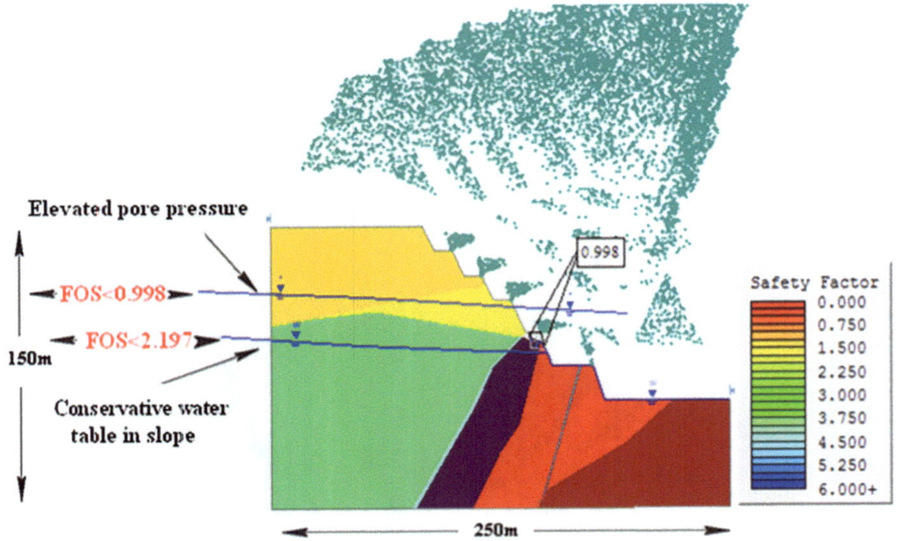

Fig. 13. Final failure model of overall slope from back–analysis using adjusted piezometric surface

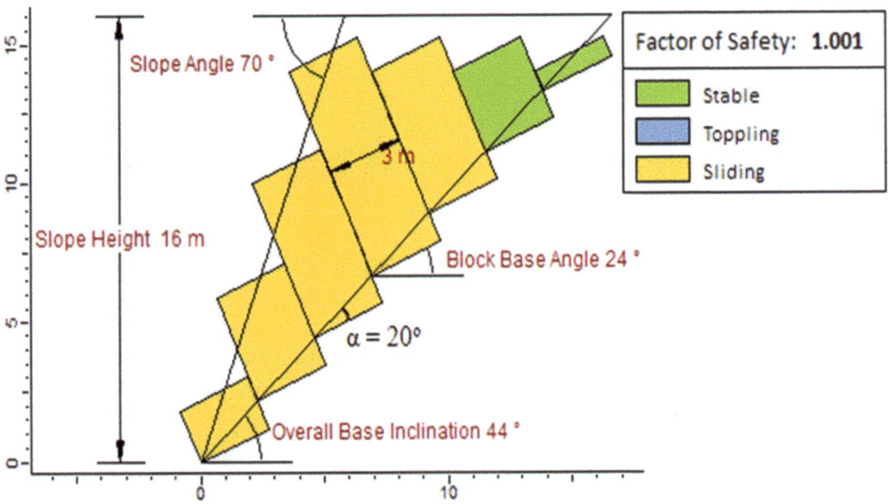

Fig. 14. Predicted toppling mechanism for the value of α =20°, which is considered for the lowest slope factor of safety 1.0

levels of displacement occurs in the partly saturated slope, and peak displacement occurs at the ramp level of weathered Magazine Shale. However, the models show that the slope was generally stable on other benches. In the comparison between bench–scale instability interpreted from the analysis using the generalised Hoek–Brown failure

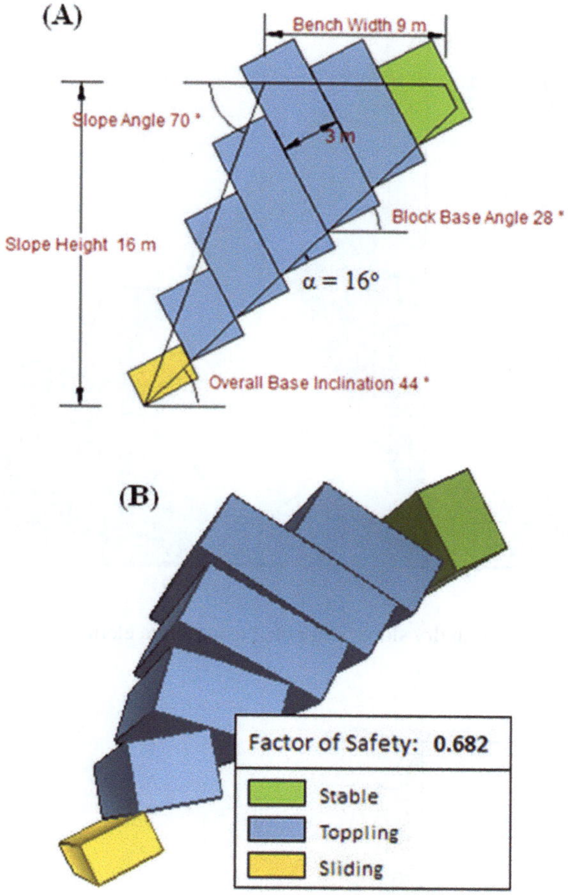

Fig. 15. Predicted toppling mechanism for the value of $\alpha = 16^{\circ}$ representing the condition of the bench slope at the time of failure. The analysis is considered for the lowest slope factor of safety 0.682, (A) 2D slope of stable block (green), toppling blocks (blue) and sliding block (yellow), (B) 3D of slope blocks during the toppling failure

criterion with the slopes modelled using the Mohr–Coulomb criterion, the later showed more stability, for it requires higher major principal stress and lower minor principal stress levels to initiate critical failure surface.

4.5 Kinematic Analyses

Analysis was conducted around the failure zone using Dips 6.0 (Rocscience 2014) to investigate the potential kinematic behaviour of critical structures (discontinuities) in the slope such as joints, bedding, faults and shears. These minor structures can control the bench–scale failures, which are more common at the Handlebar Hill pit west slope. The evaluation of the stability of geological structures is usually done through using stereographic projection, in which the discrete structures with similar orientation are grouped

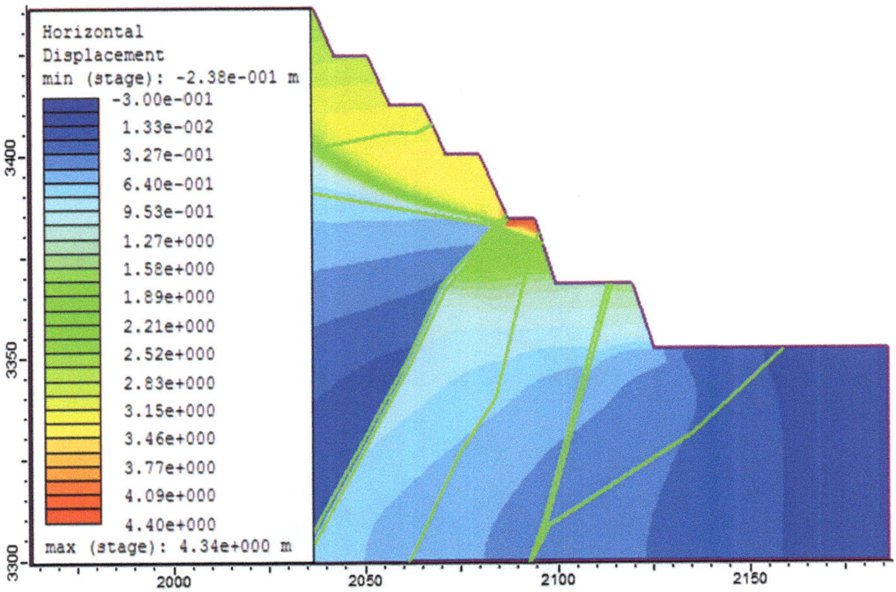

Fig. 16. Total displacement in dry slope, computed using finite element shear strength reduction method

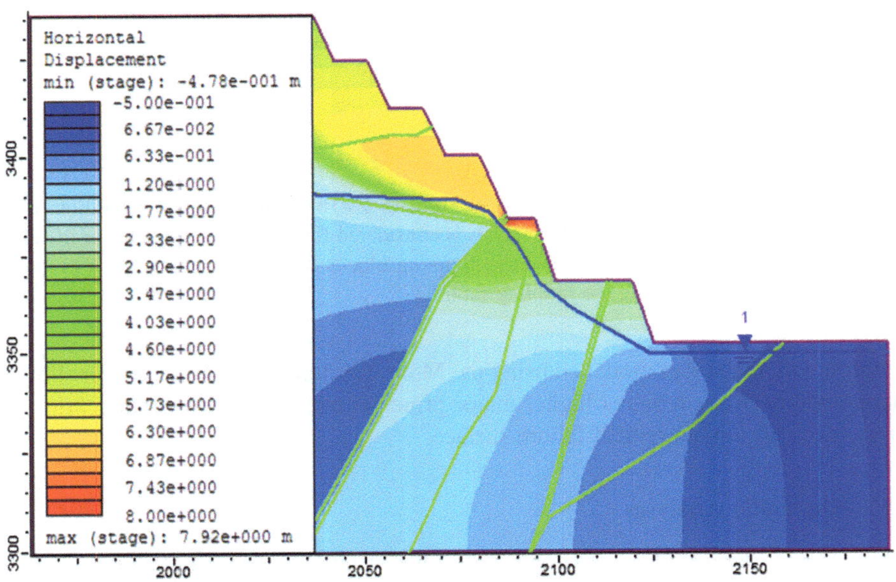

Fig. 17. Total displacement in partly saturated slope, computed using finite element shear strength reduction method

together in joint sets using various statistical techniques (Sjoberg 1999). Slope failure can occur where the potential failure plane dips at the inclination angle higher than its internal friction angle and lower than the slope face angle, thus providing a rapid and robust method for determining the stability of cut slopes (Oztekinm et al. 2006).

Initially, contours of the discontinuity data were used to locate the joint poles in the zone of failure, revealing two main joint sets at 55°/75° and 67°/288°, respectively as illustrated in Fig. 18.

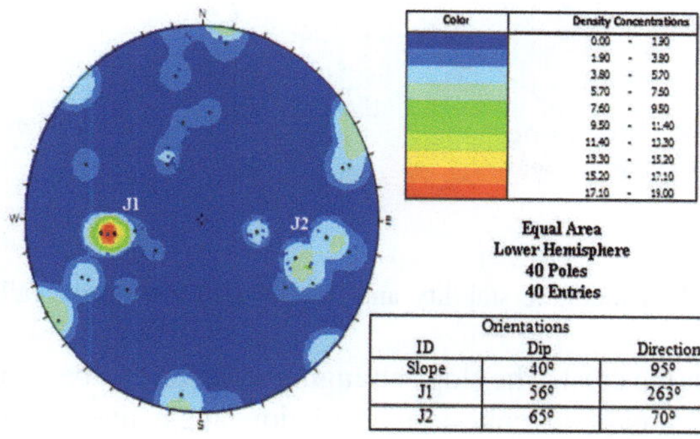

Orientations		
ID	Dip	Direction
Slope	40°	95°
J1	56°	263°
J2	65°	70°

Fig. 18. Pole frequency diagram and orientations of the discontinuities of the area around the failure

The stereographic projection of the potential failure planes in relation to the slope face highlight intersections that fall in the critical region representing the risk of developing toppling blocks (Fig. 19). These intersections are dipping into the slope and within the lateral limits of toppling mode of rock failure. The peak orientation of the dominant discontinuity sets within the slope is 50° (dip angle) and 95° (dip direction) of strike N5°E. This main discontinuity set controls the development of the slope instability in the failure zone.

5 Discussion of the Back Analysis

From the back analysis, it is clear that toppling played a critical role in this particular failure, even though the initial failure was a standard circular failure through weak rock. The back analysis using stereographic analysis (Fig. 19) and the limit equilibrium analysis (Fig. 15) both implicate toppling as a key mechanism, due to the orientation of the discontinuities and the low FOS for toppling. By comparison, the potential for larger circular failure was shown to be less likely through the back analysis using limit equilibrium (Fig. 12) and FEM (Fig. 17), both of which indicate a higher FOS than expected.

The initial failure mechanism at RL3,400 was shearing along a circular path through the upper Magazine Shale in combination with structurally controlled failures of pre-existing minor joint sets. The expansion of the opening between the Magazine Shale and the Paroo Fault, at which the circular arc was initially recognised, was then coalesced with secondary blocky toppling of the underlying materials along steeply dipping joints. The trigger remains uncertain, but is related to the local reduction of materials strength

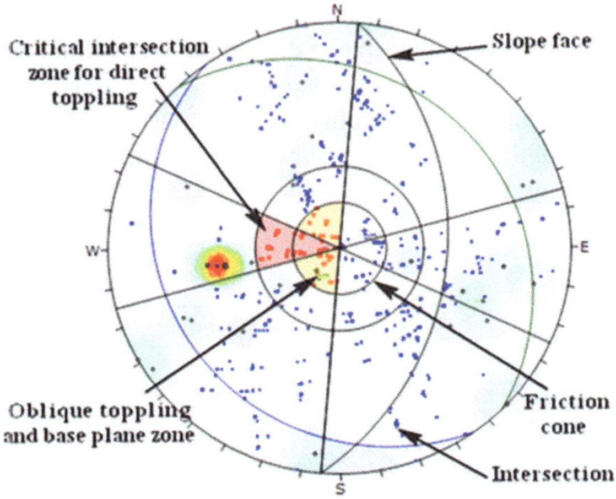

Fig. 19. Kinematic stability analyses of area around the failure

of the rock mass near the crest. The slope strength reduction can be, among other factors, a function of the decrease in cohesion or friction angle due to the propagation of fracturing. Rosengren and Associates (2007) reported that the failure was facilitated by the high degree of fracturing, with clay coated fractures, within the Magazine Shale, a relatively thin unit against the Paroo Fault. Moreover, circular failure in a section of a slope with tensional fractures is considered to be associated with long term ductile deformation of weak rock and stress release (Woldearegay et al. 2004).

One explanation is that the slope failure involved secondary toppling, defined as overturning caused by another failure mechanism, where overturning would otherwise be unlikely to occur (Goodman and Bray 1976). The weakest layer on the slope previously exhibited plastic behavior according to monitoring data, and the failure of the weakest weathered zone, adjacent to the Paroo Fault clay gouge at the top of the slope triggered the passive toppling failure as shown in Fig. 20. Since the persistence of the joints dipping into the slope is around 7 m, this model provides a credible explanation of the failure mechanism. In other studies, toppling near the slope crest was also observed resulting from circular failure that initiated at the upper slope (Wyllie and Munn 1979).

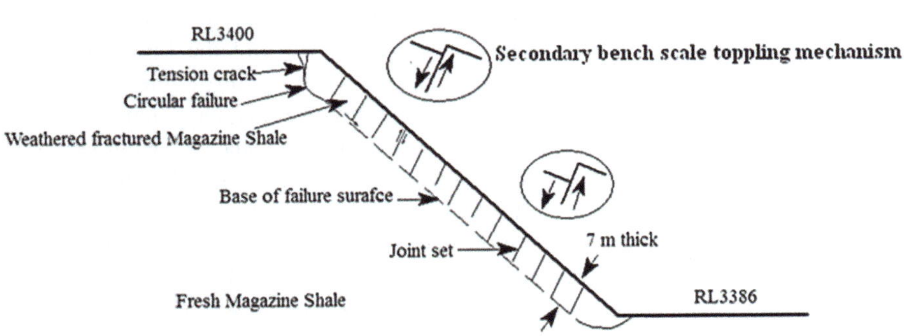

Fig. 20. Sketch of the proposed failure model of the west slope at pit

6 Conclusions

Back–analysis of the rock failure at Handlebar Hill open pit using a variety of standard models provided insights into the mechanism of slope failure and stability analysis. The factor of safety analysis showed higher than expected value of >2.0 for the slope before failure. The modeling revealed that the reduction of shear strength on RL3,400 through the initial movement along the fault discontinuity initiated the circular slope failure. The influence of rock mass properties in the leached Magazine Shale contributed the curved failure surfaces at the top of the slope. This in turn initiated secondary block toppling, and the progressive nature of the slope failure mechanisms. The failure zone is generally highly fractured rock with low uniaxial compressive strength value.

This paper provided Glencore zinc with highlight and recommendations for the failure slope based on the geotechnical and hydrogeological models developed from the back analyses. It was anticipated that the circular failure surface intersected the estimated position in the leached Magazine Shale at the top of the slope, but that slumping on the surface then coped by block toppling.

Rock slope stability monitoring in open pit mine is essential to check if the slopes are safe and rock movements are in satisfactory limits. Comprehensive radar and laser monitoring systems are being applied on the east and the west stations frequently. Therefore, collected data helps to evaluate slopes stability and displacement record can assistance to predict the behaviour of possibly instable rocks.

Acknowledgement. The authors would like to express their appreciation to Glencore zinc for providing permission to carry out this research and publish this paper. In addition, the authors would like to acknowledge Dr. Ahmed Soliman Principal Geotechnical Advisor for his support with technical reports.

References

Adhikary, D.P., Dyskin, A.V., Jewell, R.J., Stewart, D.P.: A study of the mechanism of flexural toppling failure of rock slopes. Rock Mech. Rock Eng. Springer, Heidelberg (1997). https://doi.org/10.1007/bf010-201-26

Alejano, L., Gomez–Marques, I., Matrinez–Alegria, R.: Analysis of a complex toppling–circular slope failure. Eng. Geol. ResearchGate (2010). https://doi.org/10.3390/app71-010-08

Bishop, A.W.: The use of the slip circle in the stability analysis of slopes. Geotechnology (1955). https://doi.org/10.1680/geot.1955.5.1.7

Dips 6.0.: Rocscience software, version 6.0, graphical & statistical analysis of orientation data. Math. Geosci. ResearchGate (2014). https://doi.org/10.1007/s11004-016-9635-0

Duncan, J.: State of the art: Limit equilibrium and finite–element analysis of slopes. J. Geotech. Eng. ResearchGate (1996). https://doi.org/10.1061/(asce)0733-9410

Evans, S.: An analysis of secondary toppling rock failures – the stress redistribution method. J. Eng. Geol. Geo. ScienceWorld (1981). https://doi.org/10.1144/gsl.qjeg.1981.014.02.01

Goodman, R.E., Bray, J.W.: Toppling of rock slopes. In: Proceeding of Specialty Conference on Rock Engineering for Foundations and Slopes (1976). https://doi.org/10.1007/s00603-012-0220-7

Hantz, D., Arcamon, J., Laversanne, J.: Open–pit slopes back–analysis in complex rock formation. In: Chengxiang, L., Ling, Y. (eds.) Proceedings of the International Symposium on Engineering in Complex Rock. Science Press (1988). https://doi.org/10.1007/s10346-010-0221-0
Hoek, E., Bray, W.J.: Rock slope engineering, 3rd edn. Taylor & Francis (1981). https://doi.org/10.1007/s12594-016-0500-z
Janbu, N.: Application of composite slip circles for stability analysis. In: Proceeding of European Conference on Stability of Earth Slope (1954). https://doi.org/10.1016/j.cageo.2011.10.024
Mohtrami, E., Jafari, A., Amini, A.: Stability analysis of slopes against combined circular–toppling failure. Int. J. Rock Mech. Min. Sci. (2014). https://doi.org/10.1016/j.ijrmms.2014.06.008
Morgenstern, N.R, Price, V.E.: The analysis of the stability of generalized slip surfaces. Int. J. Rock Mech. Min. Sci. ResearchGate (1965). https://doi.org/10.1080/17486025.2011.578674
Oztekin, B., Topal, T., Kolta, C.: Assessment of degradation and stability of a cut slope in limestone, Ankara–Turkey. Eng. Geol. ResearchGate (2006). https://doi.org/10.1016/j.enggeo.2005.11.012
Phase^2 V9.0/Rocsceince.: A two−dimensional finite element analysis of soil and rock for underground and excavated slopes, User's manual, Rocscience INC. Springer, Heidelberg (2014)
Priest, S.D., Brown, E.T.: Probabilistic stability analysis of variable rock slopes. Trans. Inst. Min. Metall. Elsevier (1983). https://doi.org/10.1016/0148-9062(90)94333-O
Read, J., Stacey, P.: Guidelines for Open–Pit Slope Design, p. 512. CSIRO Publishing (2009)
RocLab 1.0/Rocscience: Program for determining rock mass strength parameters, Shear strength of rock discontinuities, Rocscience Inc. Toronto, Ontario, Canada (2014)
Rosengren, K., Associates: Proposed Handlebar Hill open cut geotechnical review (Technical Report No. 27010). Xstrata, Mt Isa, Australia (2007)
Rosengren, K., Associates.: Handlebar Hill open cut Pit. Internal technical report, in Xstrata report in 2011, Xstrata, Mt Isa, Australia (2008)
Sjoberg, J.: Analysis of large scale rock slopes. Doctoral Thesis, Lulea University of Technology, Lulea, Sweden (1999)
Skempton, A.W.: The ($\varphi = 0$) analysis of stability and its theoretical basis. In: Proceeding of the 2nd International Conference of soil Mechanics and Foundation Engineering. Ice Virtual Library (1948). https://doi.org/10.1680/geot.1948.1.1.1
Slide 6.0.: 2D limit equilibrium slope stability analysis. User's manual, Rocscience. *nrc research press* (2014). https://doi.org/10.1139/cgj-2014-0111
Spencer, E.: A method of analysis of the stability of embankments assuming parallel inter–slice forces. J. Geotech. Sci. Educ. (1967). https://doi.org/10.12691/ajna-1-1-2
Terzaghi, K.: Theoretical Soil Mechanics, 1st edn. Wiley (1943). https://doi.org/10.1002/9780470172766
Warren, N.S., Watters, J.R.: Limit equilibrium stability analysis utilizing geotechnical date at Mt. Baker. ProQuest Dissertations Publishing, University of Nevada, Reno (2008). https://doi.org/10.1139/cgj-2014-0111
Woldearegay, K., Riedmuller, G., Schubert, W., Mogessie, A.: Controlling parameters and failure mechanisms of a large scale landslide in Ethiopia. FELSBAU **22**(3) (2004)
Wyllie, D.C., Munn, F.J.: The use of movement monitoring to minimize production losses due to pit slope failures. In: Proceedings, 1st International Symposium on Stability in Coal Mining (1979)
Wyllie, D.C., Mah, C.W.: Rock Slope Engineering, Civil and Mining, 4th edn. Routledge (2004)

Reflection of Processes of Non-equilibrium and Two-Phase Filtration in Fluid Saturated Hierarchic Inclusion in a Block Layered Medium by Data of Active Wave Geophysical Monitoring

Olga Hachay[1(✉)], Andrey Khachay[2], and Oleg Khachay[2]

[1] Institute of Geophysics Ural Branch of Russian Academy of Sciences, Yekaterinburg, Russia
olgakhachay@yandex.ru
[2] Ural Federal University, Yekaterinburg, Russia

Abstract. A new method for modeling acoustic monitoring of a layered-block elastic medium with several inclusions of various physical-mechanical and phase hierarchical structures was developed. An iterative process for solving the direct problem is developed for the case of three hierarchical inclusions of l, m, s-th ranks on the basis of the use of 2D integral and integral- differential equations. The degree of hierarchy of inclusions is determined by the values of their ranks, which can be different for each inclusion. Hierarchical inclusions are located in different layers above each other: the upper is anomalously strained, the second is fluid-saturated and the third is anomalously dense. The degree of filling inclusions of each rank for all three hierarchical inclusions is different. The results of the simulation can be used for monitoring of the fluid flow of oil deposits.

Keywords: Hierarchical environment · Acoustic field · Iterative algorithm
Integral and integro-differential equations · Direct problem

1 Introduction

The most important result of geomechanical and geodynamic studies of the past century was the discovery of a close relationship between global geodynamic and local geomechanical processes caused by mining operations, especially in tectonically active zones. No less important result of the research was the conclusion about the fundamental role of the block-hierarchical structure of rocks and massifs for explaining the existence of a wide range of nonlinear geomechanical effects and the emergence of complex self-organizing geosystems. Hierarchical structure is typical for many systems, especially for the Earth's lithosphere, where more than 30 hierarchical levels from tectonic plates with a length of thousands of kilometers to individual mill grains of millimeter size were identified by geophysical studies (Prangishvilli et al. 2001). Thus, the earth's crust is not a continuous medium, but a discrete block system and,

J. Wasowski and T. Dijkstra (Eds.): GeoMEast 2018, SUCI, pp. 31–38, 2019.
https://doi.org/10.1007/978-3-030-02032-3_3

like any synergetic discrete ensemble, has hierarchical and self-similar properties (Kocharjan et al. 2003).

Possible reduction in economic indicators in the global oil and gas industry is due to outdated and largely erroneous concepts on the structure of non-oil and gas fields in the geological section. We can note several negative consequences in the economy of the fuel and energy complex, caused by outdated and erroneous ideas. First, unjustifiably large volumes of drilling wells in those sections of the fields where either there are no deposits, or there are very poor deposits, the development of which is knowingly unprofitable. Secondly, the various methods of influencing the reservoir with the aim of increasing oil recovery in those areas where there is practically no oil only worsen the economic indicators with erroneous ideas about the genesis and spatial structure of the deposits. It is known that the volumes of injected liquid grow several times faster than the volumes of incremental oil production. Thirdly, unreasonably high expenses for drilling and pumping water not only reduce the overall profitability of hydrocarbon production, but also significantly increase the level of environmental pollution, the liquidation of which will require additional costs. Fourth, the erroneous concepts of the origin and evolution of hydrocarbon deposits have led to such production technologies, under which there is a very rapid depletion of the deposits and significantly reduced the renewability of resources. All this unjustifiably quickly reduces economic indicators in developed fields. The main elements of the new paradigm in the fuel and energy complex are the search, development and technology of development of small-sized, but extraordinarily rich in reserves of active oil and gas generation. The geosoliton concept of the formation of hydrocarbons and the formation of large-scale, but rich in reservoirs, is the main conceptual basis of the new paradigm (Bembel et al. 2004). To date, new effective technologies can be attributed the results published in the paper Levashov et al. (2017).

The processes of development of oil and gas fields are associated with the motion of multiphase multicomponent media that are characterized by no equilibrium and nonlinear rheological properties. The real behavior of reservoir systems is determined by the complexity of the rheology of moving fluids and the morphological structure of the porous medium, as well as by the variety of processes of interaction between the liquid and the porous medium (Khasanov et al. 2003). Accounting for these factors is necessary for a meaningful description of filtration processes due to the nonlinearity, nonequilibrium and heterogeneity inherent in real systems. In this case, new synergetic effects are revealed (loss of stability with the appearance of oscillations, the formation of ordered structures). This allows us to propose new methods for monitoring and managing complex natural systems that are tuned to account for these phenomena. Thus, the reservoir system from which to extract oil is a complex dynamic hierarchical system.

When constructing a mathematical model of a real object, it is necessary to use, as a priori information, active and passive monitoring data obtained during the current operation of the facility. In papers of Hachay et al. (2011), (2013), modeling algorithms were constructed for 3-D heterogeneities in electromagnetic case, for 2-D heterogeneities in the seismic case and for an arbitrary type of excitation source of an N-layer medium with hierarchical elastic inclusions located in the J-th layer. In the paper of Hachay et al. (2014), a new 2D modeling algorithm for sound diffraction on elastic and porous, fluid-saturated inclusions of hierarchical structure located in the J-th layer of an N-layer elastic medium was developed. In Hachay et al. (2017) modeling algorithms were constructed in the acoustic case for a 2-D heterogeneity and an arbitrary type of excitation source of an N-layer medium with a separate hierarchical anomaly density, stresses and plastic inclusions located in the J-th layer.

In this paper, using the method described in Hachay et al. (2011), (2013), (2014), (2017), an algorithm for modeling the acoustic field (longitudinal acoustic wave) has been developed in the form of an iterative process for solving a direct problem for the case of three hierarchical inclusions of *l, m, s*-ranks using 2D integral and integro-differential equations. The degree of hierarchy of inclusions is determined by the values of their ranks, which can be different. Hierarchical inclusions are located in different layers above each other: the top is anomalously stressed, the second is by fluid saturated and the third is anomaly dense (Fig. 1).

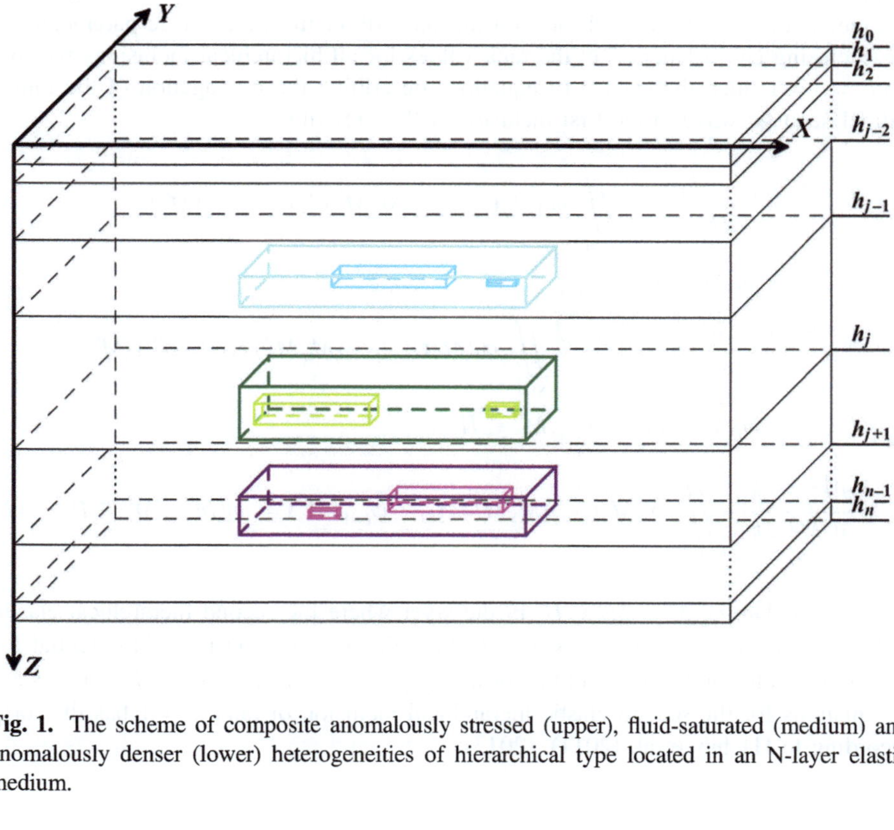

Fig. 1. The scheme of composite anomalously stressed (upper), fluid-saturated (medium) and anomalously denser (lower) heterogeneities of hierarchical type located in an N-layer elastic medium.

1.1 Algorithm of Modeling of Sound Diffraction on 2D Composite Heterogeneity of Hierarchical Type, Located in N-Layered Elastic Environment

An algorithm for modeling of sound diffraction on a two-dimensional elastic hierarchical inclusions located in the J-th layer of an N-layer medium is described in Hachay et al. (2013), (2017). $G_{Sp,j}(M, M^0)$- the source function of the seismic field, the boundary value problem for which was formulated in Khachay (2006a, b), $k_{1ji}^2 = \omega^2(\sigma_{ji}/\lambda_{ji})$ is the wave number for the longitudinal wave; in the above expression, the index ji denotes the property of the medium inside the heterogeneity, ja denotes the property of the medium outside the heterogeneity, λ is the Lamé constant, σ -density of medium, ω - circular frequency, $\vec{u} = grad\ \varphi$- vector of displacements, φ^0- potential of normal seismic field in layered medium in the absence of heterogeneity: $\varphi_{ji}^0 = \varphi_{ja}^0$. φ_l^0- potential of a normal seismic field in a layered medium in the absence of a heterogeneity of the previous rank l if $l = 2 \ldots L$, if $l = 1$, which coincides with the corresponding expression (Hachay et al. 2017). We assume that the density of the first hierarchical inclusion for all ranks l and the enclosing layer are identical, and elastic parameters of hierarchical inclusion for all ranks differ from the elastic parameters of the enclosing environment. Let the rank values for all hierarchical inclusions will be: $l = m = s = 1$, then the system of equations describing the propagation of the longitudinal acoustic wave in the first inclusion will be rewritten as:

$$\begin{aligned} &\frac{(k_{1(j-1)il}^2 - k_{1(j-1)}^2)}{2\pi} \iint\limits_{S1cl} \varphi_l(M) G_{Sp,(j-1)}(M, M^0) d\tau_M + \varphi_{l-1}^0(M^0) \\ &= \varphi_l(M^0),\ M^0 \in S_{1Cl} \\ &\frac{\sigma_{(j-1)il}(k_{1(j-1)il}^2 - k_{1(j-1)}^2)}{\sigma(M^0)2\pi} \iint\limits_{S1cl} \varphi_l(M) G_{Sp,(j-1)}(M, M^0) d\tau_M + \varphi_{l-1}^0(M^0) \\ &= \varphi_l(M^0),\ M^0 \notin S_{1Cl}, M^0 \in \Pi_{j-1} \end{aligned} \tag{1}$$

$$\frac{\sigma_{jil}(k_{1jil}^2 - k_{1j}^2)}{\sigma(M^0)2\pi} \iint\limits_{S1cl} \varphi_l(M) G_{Sp,j}(M, M^0) d\tau_M + \varphi_{l-1}^0(M^0) = \varphi_l(M^0),\ M^0 \in \Pi_j \tag{2}$$

We calculate $\varphi_l(M^0)$, $M^0 \in \Pi_j$ in the layer where the second hierarchical elastic inclusion is located, using expression (2), then the normal acoustic field potential for the second inclusion is written in the form: $\varphi_{m-1}^0(M^0) = \varphi_l(M^0), M^0 \in \Pi_j$. The system of equations for the second elastic hierarchical inclusion of rank $m = 1$ has the form according to Hachay et al. (2011), (2013):

$$\begin{aligned}
&\frac{(k_{1jim}^2 - k_{1j}^2)}{2\pi} \iint\limits_{S_{2Cm}} \varphi_m(M) G_{Sp,j}(M, M^0) d\tau_M + \frac{\sigma_{ja}}{\sigma_{jim}} \varphi_{m-1}^0(M^0) \\
&- \frac{(\sigma_{ja} - \sigma_{jim})}{\sigma_{jim} 2\pi} \int\limits_{C_{2m}} G_{Sp,j} \frac{\partial \varphi_m}{\partial n} dc = \varphi_m(M^0), M^0 \in S_{2Cm}; \\
&\frac{\sigma_{jim}(k_{1jim}^2 - k_{1j}^2)}{\sigma(M^0) 2\pi} \iint\limits_{S_{2Cm}} \varphi_m(M) G_{Sp,j}(M, M^0) d\tau_M + \varphi_{m-1}^0(M^0) \\
&- \frac{(\sigma_{ja} - \sigma_{jim})}{\sigma(M^0) 2\pi} \int\limits_{C_{2m}} G_{Sp,j} \frac{\partial \varphi_m}{\partial n} dc = \varphi_m(M^0), M^0 \notin S_{2Cm};
\end{aligned} \tag{3}$$

$$\begin{aligned}
&\frac{\sigma_{(j+1)im}(k_{1(j+1)im}^2 - k_{1(j+1)}^2)}{\sigma(M^0) 2\pi} \iint\limits_{S_{2Cm}} \varphi_m(M) G_{Sp,(j+1)}(M, M^0) d\tau_M + \varphi_{m-1}^0(M^0) \\
&- \frac{(\sigma_{(j+1)a} - \sigma_{(j+1)im})}{\sigma(M^0) 2\pi} \int\limits_{C_{2m}} G_{Sp,(j+1)} \frac{\partial \varphi_m}{\partial n} dc = \varphi_m(M^0), M^0 \notin S_{2Cm}, \in \Pi_{j+1};
\end{aligned} \tag{4}$$

We calculate $\varphi_m(M^0), M^0 \notin S_{2Cm}, \in \Pi_{j+1}$ in the layer where the third hierarchical anomalous density inclusion is located using expression (4), then the normal potential of the acoustic field is: $\varphi_{s-1}^0(M^0) = \varphi_m(M^0), M^0 \in \Pi_{j+1}$. We assume that the elastic parameters of the third hierarchical inclusion for all ranks s and the enclosing layer are the same, and the density of the hierarchical inclusion for all ranks differs from the density of the host environment, then the system of equations for the third hierarchical inclusion of rank $s = 1$ has the form according to Hachay et al. (2017):

$$\begin{aligned}
&\frac{(k_{1(j+1)is}^2 - k_{1(j+1)}^2)}{2\pi} \iint\limits_{S_{3Cs}} \varphi_s(M) G_{Sp,(j+1)}(M, M^0) d\tau_M + \frac{\sigma_{(j+1)a}}{\sigma_{(j+1)is}} \varphi_{s-1}^0(M^0) \\
&- \frac{(\sigma_{(j+1)a} - \sigma_{(j+1)is})}{\sigma_{(j+1)is} 2\pi} \int\limits_{C_{3s}} G_{Sp,(j+1)} \frac{\partial \varphi_s}{\partial n} dc = \varphi_s(M^0), M^0 \in S_{3Cs}; \\
&\frac{\sigma_{(j+1)is}(k_{1(j+1)is}^2 - k_{1(j+1)}^2)}{\sigma(M^0) 2\pi} \iint\limits_{S_{3Cs}} \varphi_s(M) G_{Sp,(j+1)}(M, M^0) d\tau_M + \varphi_{s-1}^0(M^0) \\
&- \frac{(\sigma_{(j+1)a} - \sigma_{(j+1)is})}{\sigma(M^0) 2\pi} \int\limits_{C_m} G_{Sp,(j+1)} \frac{\partial \varphi_s}{\partial n} dc = \varphi_s(M^0), M^0 \notin S_{3Cs} \in \Pi_{j+1};
\end{aligned} \tag{5}$$

$G_{Sp,(j+1)}(M, M^0)$- function of the source of the seismic field, it coincides with the function (Khachay 2006a, b), $k_{1(j+1)is}^2 = \omega^2(\sigma_{(j+1)is}/(\lambda_{(j+1)is});\ \lambda_{(j+1)is} = \lambda_{(j+1)a}$; - wave number for the longitudinal wave and elastic parameters, in the above expression, the index ji denotes the belonging of the medium properties inside the heterogeneity, ja - outside the heterogeneity, $s = 1, \ldots, S$ - number of the hierarchical levels, $\varphi^0_{s-1}(M^0) = \varphi_m(M^0), M^0 \in \Pi_{j+1}$ is the potential of the normal acoustic field in the layer $j + 1$ in the absence of the third heterogeneity of the previous rank. We calculate $\varphi_s(M^0), M^0 \notin S_{3Cs} \in \Pi_{j-1}$ in the layer $j - 1$ using expression (6):

$$\begin{aligned}
&\frac{\sigma_{(j-1)is}(k^2_{1(j-1)is} - k^2_{1(j-1)})}{\sigma(M^0)2\pi} \iint\limits_{S_{3Cs}} \varphi_s(M) G_{Sp,(j-1)}(M, M^0) d\tau_M + \varphi^0_{s-1}(M^0) \\
&- \frac{(\sigma_{(j-1)a} - \sigma_{(j-1)is})}{\sigma(M^0)2\pi} \int\limits_{C_m} G_{Sp,(j-1)} \frac{\partial \varphi_s}{\partial n} dc = \varphi_s(M^0), M^0 \notin S_{3Cs} \in \Pi_{j-1};
\end{aligned} \tag{6}$$

The values of L, M, and S are the maximum values of the ranks of the hierarchy for the three inclusions. In this paper,

$$L = 3, M = 3, S = 4. \tag{6'}$$

Let: $l = l + 1$; $m = m + 1$; $s = s + 1$. If $l < 3$ or $l = 3$, then we turn to the algorithm (1)–(6). If $l > 3$, and $m = 2$, then we calculate $\varphi_s(M^0), M^0 \notin S_{3Cs} \in \Pi_j$ in the layer j using expression (7):

$$\begin{aligned}
&\frac{\sigma_{jis}(k^2_{1jis} - k^2_{1j})}{\sigma(M^0)2\pi} \iint\limits_{S_{3Cs}} \varphi_s(M) G_{Sp,j}(M, M^0) d\tau_M + \varphi^0_{s-1}(M^0) \\
&- \frac{(\sigma_{ja} - \sigma_{jis})}{\sigma(M^0)2\pi} \int\limits_{C_m} G_{Sp,(j-1)} \frac{\partial \varphi_s}{\partial n} dc = \varphi_s(M^0), M^0 \notin S_{3Cs} \in \Pi_j;
\end{aligned} \tag{7}$$

$\varphi^0_{m-1}(M^0) = \varphi_{s-1}(M^0), M^0 \in \Pi_j$ and we proceed to the algorithm (3)–(6), if $m = 3$, $\varphi^0_{m-1}(M^0) = \varphi_{s-1}(M^0), M^0 \in \Pi_j$ then we turn to the algorithm (3)–(6). If $m > 3$, and $s < 4$ or $s = 4$, $\varphi^0_{s-1}(M^0) = \varphi_{s-1}(M^0), M^0 \in \Pi_{j+1}$, then we go to (5)–(6). If $s > 4$, then we pass to (7′).

$$\begin{aligned}
&\frac{(k^2_{1jim} - k^2_{1j})}{2\pi} \iint\limits_{S_{2Cm}} \varphi_m(M) G_{Sp,j}(M, M^0) d\tau_M + \frac{\sigma_{ja}}{\sigma_{jim}} \varphi^0_{m-1}(M^0) \\
&- \frac{(\sigma_{ja} - \sigma_{jim})}{\sigma_{jim}2\pi} \int\limits_{C_{2m}} G_{Sp,j} \frac{\partial \varphi_m}{\partial n} dc = (\varphi_m(M^0) + \alpha p_2), M^0 \in S_{2Cm}; \\
&\frac{\sigma_{jim}(k^2_{1jim} - k^2_{1j})}{\sigma(M^0)2\pi} \iint\limits_{S_{2Cm}} \varphi_m(M) G_{Sp,j}(M, M^0) d\tau_M + \varphi^0_{m-1}(M^0) \\
&- \frac{(\sigma_{ja} - \sigma_{jim})}{\sigma(M^0)2\pi} \int\limits_{C_{2m}} G_{Sp,j} \frac{\partial \varphi_m}{\partial n} dc = \varphi_m(M^0), M^0 \notin S_{2Cm};
\end{aligned} \tag{7'}$$

where $\alpha = 1 - \chi - \frac{K}{K_0}$, $K = \lambda$-complete compression module, χ-porosity, K_0-the true modulus of phase compressibility, pore hydrostatic pressure $p2$. The first equation of system (3) is rewritten in the form in accordance with the fact that in the second inclusion the heterogeneities of the second rank become fluid or oil-saturated. Then go to (5)–(6).

$$\frac{\sigma_{jis}(k_{1jis}^2 - k_{1j}^2)}{\sigma(M^0)2\pi} \iint\limits_{S_{3Cs}} \varphi_s(M) G_{Sp,j}(M, M^0) d\tau_M + \phi_{s-1}^0(M^0)$$
$$- \frac{(\sigma_{ja} - \sigma_{jis})}{\sigma(M^0)2\pi} \int\limits_{C_m} G_{Sp,j} \frac{\partial \varphi_s}{\partial n} dc = \varphi_s(M^0), M^0 \notin S_{3Cs} \in \Pi_j; \tag{8}$$

We calculate $\varphi_s(M^0), M^0 \notin S_{3Cs} \in \Pi_j$ in all the layers $j = 1, \ldots n$ with the use of the expression (8). The algorithm stops if the hierarchy ranks become larger than the given numbers (6′). If at some hierarchical level the structure of the local heterogeneity breaks down into several heterogeneities, then the double and contour integrals in the expressions (1–7′) are taken over all heterogeneities of a given rank.

2 Conclusions

Iterative modeling algorithms are constructed in the seismic case in the acoustic approximation for a composite hierarchical heterogeneity. When constructing a complex seismic gravity model without taking into account the anomalous influence of a stress-strain state within a hierarchical inclusion, called the cushion of the deposit, an analysis of the anomalous acoustic effect using data on the propagation of a longitudinal wave shows that the influence of anomalous elastic parameters in the seismic model cannot be neglected, since they affect on the values of the unknown anomalous densities. If these values are used in constructing a density gravitational model without taking into account the effect of elastic parameters, these density values will not reflect the material composition of the analyzed medium. When constructing an anomalously strained geomechanical model without taking into account the anomalous influence of density heterogeneities inside the hierarchical inclusion, which is the substrate for a two-phase deposit, the values of the unknown anomalous elastic parameters that cause the anomalous stress state in the cushion using data on propagation of the transverse wave will be determined incorrectly. These values of elastic parameters will not reflect the real stress state of the analyzed medium over the fluid containing the deposit, which in turn is represented by a hierarchical multi-neighborhood environment. For the first time, the proposed iterative algorithm for modeling a hierarchically complex two-phase medium can be used to manage the production of viscous oil in mine conditions and light oil in sub horizontal wells (Khachay et al. 2016). The benefits of the new paradigm, of the new methods of oil and gas mapping and monitoring using geophysical wave fields on the base of developed mathematical algorithms can be seriously enlarged.

References

Bembel, R.M., et al.: Geosoliton: The Functional System of the Earth, the Concept of Exploration and Development of Hydrocarbon Deposits. Vector Book, Tyumen (2004)

Hachay, O.A., et al.: Integration of seismic and electromagnetic active methods for mapping and monitoring the state of two-dimensional heterogeneities in an N-layer medium. Bull. South Ural State Univ. Ser. Comput. Technol. Manage. Radio Electr. **2**, 49–56 (2011)

Hachay, O.A., et al.: Modeling of electromagnetic and seismic fields in hierarchically heterogeneous media. In: Bull. South Ural. State University. Ser. Comput. Math. Inform. **2** (2), 48–55 (2013)

Hachay, O.A., et al.: Reflection of no equilibrium two-phase filtration processes in oil saturated hierarchical media in active wave geophysical monitoring data. Min. Inf. Anal. Bull. **4**, 232–238 (2014)

Hachay, O.A., et al.: Integration of acoustic, gravitational, and geomechanical fields in hierarchical environments. Min. Inf. Anal. Bull. **4**, 328–336 (2017)

Khachay, A.Y.: Algorithm for solving the direct dynamic seismic problem when excited by a horizontal point force located in an arbitrary layer of an n-layer elastic isotropic medium. In: Informatics and Mathematical Modeling, pp. 170–278. Ural State University, Ekaterinburg (2006a)

Khachay, A.Y.: Algorithm for solving the direct dynamic seismic problem when excited by a vertical point force located in an arbitrary layer of an n-layer elastic isotropic medium. In: Informatics and Mathematical Modeling, pp. 279–310. Ural State University, Ekaterinburg (2006b)

Khachay, Y., et al.: Effect of convective transport in porous media on the conductions of organic matter maturation and generation of hydrocarbons in trap rocks complexes. Energy Procedia. **74**, 79–83 (2016)

Khasanov, M.M., et al.: Nonlinear and no equilibrium effects in rheological complex environments. Institute of Computer Research, Izhevsk, Moscow (2003)

Kocharjan, G.G., et al.: Dynamics of deformation of block massif. IKZ "Akademkniga", Moscow (2003)

Levashov, S., et al.: On the possibility of using mobile and direct-prospecting geophysical technologies to assess the prospects of oil-gas content in deep horizons. In: Oil and Gas exploration. Methods and Application, pp. 209–236. AGU, Willey, USA (2017)

Prangishvilli, I.V., et al.: System laws and regularities in electrodynamics, nature and society. Nauka, Moscow (2001)

Evaluation of the Landslide in Erzurum Ski-Jumping Complex - A Case Study

Murat Mollamahmutoğlu(✉)

Civil Engineering Department, Engineering Faculty, Gazi University, Ankara, Turkey
mmolla@gazi.edu.tr

Abstract. We present the assessment of a landslide that took place in the Erzurum ski-jumping complex constructed for the 2011 Winter Universiade in Turkey. The lower part of the ski jumps collapsed. Three of the jumps were completely destroyed whilst the two larger jumps were severely damaged. To begin with, the collapsed ski-jumping complex was inspected immediately and periodically after the landslide. Thereafter, the related projects and site investigation report were obtained and examined in detail. Since the report for the site was found to be insufficient, additional in-situ and laboratory tests were planned and carried out to better understand the geotechnical characterization of the site and compensate for the absence of data needed for the related geotechnical analyses. Furthermore, the detailed slope stability analyses under different conditions regarding various factors were conducted. All these studies revealed that the landslide was not sudden but went through a creeping process. The collapse of ski-runs in the landslide was due to the inadequate geotechnical investigation and engineering judgment, design and application mistakes and the lack of stability analyses and supervision.

1 Introduction

Landslides are a worldwide problem. They can lead to loss of life, loss of livestock and damage the residential and industrial developments, villages or even entire towns, destroying agricultural and forest lands. 2620 fatal landslides between 2004 and 2010 killed a total of 32,322 people excluding landslides triggered by earthquakes (Petley 2012).

The triggering mechanisms of natural slope failures frequently comprise a complex interaction between hydrological and geotechnical processes, which in turn depends on irregular topography, hydro-geotechnical properties, boundary conditions such as permeability, and the initial state of the slope (Sorbino and Nicotera 2013). From a geotechnical perspective, one of the main reasons for slope failure is the loss of matric suction and hence a loss of effective stress as water infiltrates the soil.

Many researchers (Kim et al. 2004; Fredlund et al. 2012) have studied rainfall-induced landslides through laboratory, field tests, and numerical analysis. They have presented a mechanism for rainfall-induced landslides that commonly occur when the wetting bands progress into the soil, resulting in a loss of suction and a reduction in effective stress (Lu and Godt 2008).

J. Wasowski and T. Dijkstra (Eds.): GeoMEast 2018, SUCI, pp. 39–48, 2019.
https://doi.org/10.1007/978-3-030-02032-3_4

Geologic factors have also been found to cause mass movements on the slopes and these include the presence of shallow soils over hard, impermeable rocks or glacial till, soft, clay-rich rocks that produce thick plastic soil mantles, alignment of lithology parallel to the ground slope and planar rock structures, unconsolidated or weakly consolidated deposits (Sidle 1985).

Furthermore, landslides are commonly triggered by the modification of slope stability conditions, increasing stress, or reductions in strength (Rahardjo et al. 1988; Ng et al. 2003, Ng et al. 2008). The landslide analysis on a large scale is usually performed using deterministic methods such as the Limited Equilibrium Method (LEM) and Finite Element Method (FEM). They are also useful in determining the factors involved and the mechanism of slope failures, such as slope geometry, the mechanical properties, and equilibrium condition.

In this paper, we present the results of the comprehensive investigation aimed at determining the causes of the activation of a landslide in the Erzurum ski-jumping complex, Turkey.

2 Study Area

2.1 Description

The Erzurum Ski Jump, or officially known as Türk Telekom Ski Jumping Towers, is a ski jumping complex located on the Kiremitlik Hill at the base of Palandöken Mountain, southwest of Erzurum in eastern Turkey. The complex, completed at the end of 2010, consists of a large hill and a normal hill jumping tower, as well as three additional smaller slopes of K-65, K-40, and K-20 for training purposes and for use by young jumpers. The center with five ski jumping slopes covers an area of 30 ha. The large hill has a K-point of 125 and a hill size of 140 with a 43.47 m high jumping tower. It has a 99 m long inrun with an angle of 35°, a take-off angle of 11°, which is 7 m long and 3 m high. The slope of the landing is 34.57°. The total height is 136.8 m. The normal hill's K-point is 95 and hill size 109. It has a jumping tower of 20.98 m height. The inrun is 88 m long at an angle of 35°. The take-off length is 6.5 m, the height 2.5 m with an angle of 10.5°. The landing's angle is 34.25°. The normal hill has a total height of 107.1 m. The two Olympic-size towers have cylindrical structures at the top. The venue has 10,000 spectator capacity (Fig. 1a).

During the time when no competitions are held, the facility attracts local population and tourists for recreation activities and social events at the hotel restaurant and the restaurant atop the 43 m-high tower, which offers a panoramic view of the entire city. The whole complex cost about €20 million. The venue hosted fourth international competitions, the last one being in 2012 and then the whole ski-runs was completely damaged in a landslide in 2014 (Fig. 1b).

2.2 Regional Geology and Tectonic Setup

The study area is characterized by the presence of Pliocene age Gelinkaya Formation deposited in lake and river environment. It is mainly composed of less-consolidated siltstone (Fig. 2a) sporadically interbedded with sandstone, gravels, and tuff (Figs. 2b, 3a, and b).

Fig. 1. Views of the completed and failed ski jump complex

Fig. 2. Close-up views of thinly layered siltstone and siltstone with coarse poorly sorted materials

The strata in the Kiremitlik hill where the landslide occurred are highly fractured and deformed. The faults on both sides of Kiremitlik hill are thought to be the source of tectonic deformations but there is not sufficient information about them (Yılmaz and Uysal 1988).

3 Methodology

The method of this research comprised an assessment of available data, surface and subsurface investigations, in-situ and laboratory tests and the slope stability modeling and analyses.

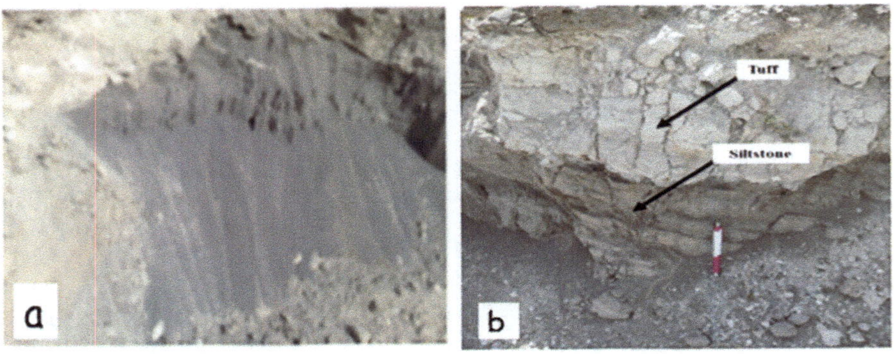

Fig. 3. Views of siltstone with less-consolidated sandstone layers and alternation of siltstone and tuff

3.1 Assessment of Available Data

Geophysical, Geological and Geotechnical Studies

Surprisingly, there were a few boreholes within the slope that failed. Its soil structure was inferred by means of some surrounding boreholes, which, however, had insufficient depth of 10 to 12 m; none of them were extended to a depth of firm strata. Similarly, the geophysical survey was also superficial reaching the depth of 20 m at most. During borehole drilling, SPT was used to evaluate the strength of the layers encountered, but the results were misleading due to the fact that most of the drilled layers contained gravels and stones. Moreover, the position and orientation of the layers in the landslide area were misinterpreted as horizontal whereas the later site investigation revealed the presence of more complex structure. Thirteen new boreholes whose depths ranged from 30 to 111 m, were driven to better understand the local soil structure. The borehole locations are shown in Fig. 4. Since the access to failed area was not possible they are scattered around wherever possible.

Related Projects

Cuts and fills were planned in the related projects to form standard slopes of ski-runs (Fig. 5a). In addition, to protect skiers from the wind effect in the course of jumping earthfill was designed next to the top of K125 and K65 ramps (Fig. 5b). But none of these arrangements were based on the slope stability analyses.

A great mistake in the planning was the missing of the effective drainage system to prevent the adverse effect of surface water infiltration into the ground (Fig. 5c).

Managerial Control and Consultancy Services

Supervision of construction activities on site was carried out jointly by the engineers of the Ministry of Youth and Sport and the project firm. Nevertheless, it seemed that they were not qualified in assessing the site investigation report and the related projects and reading the conditions encountered on the site during construction activities. Figure 6 showed that some local slope failures took place during field works.

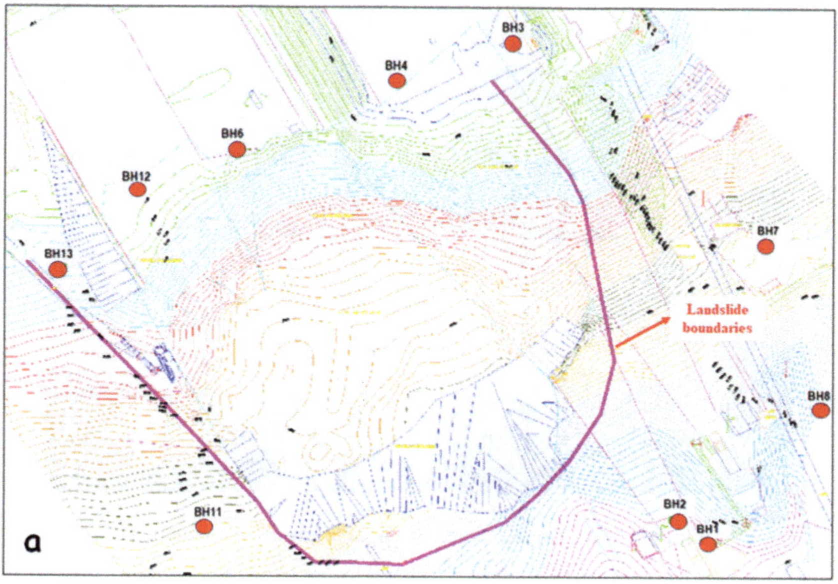

Fig. 4. Boreholes layout

Fig. 5. Slope arrangements of K125, K65, and K40 in runs

Slope Creeping

The ski jump complex in question was completed at the end of 2010 and brought into service in January 2011. After hosting the event of 2012 junior winter sport, heave process took place at the lower part of K 125 ski-run concrete slab between the 80th and 90th m towards the end of May 2012 (Fig. 7a). This was considered to be the slab damage and repaired. In the first week of June 2014, on a rainy day, at the same place of K125 ski-run concrete slab, another heave occurred extending to K65, K40, and K20 ski-runs' concrete slabs (Fig. 7b). In the meanwhile, there were cracks above the heaving line. The damaged parts were again repaired. Notwithstanding, a week later after the repairment, heave and cracks repeated on the same lines. From that day on, creeping of slopes increasingly continued. The process went on about five days and the cracks on the upper part of K125 and K95 ski-runs were widened (Fig. 8a). At the same time, on the other parts of the ski-runs undulating surfaces developed. Furthermore, the windbreaker earthfill near the top of K125 ski-run ruptured and the surface deformations increased (Fig. 8b). The area was completely evacuated a day before the failure of the ski-runs, that is on 15 July 2014.

Fig. 6. Local slope failures in the field during construction

Fig. 7. Creeping processes of K125, K65, K40 and K20 ski-runs

Fig. 8. Ruptures on the upper part of K125 ski-run and the windbreaker earthfill

It seemed that the slope deformation increments coincided with the periods of precipitation, watering, and snowing of the ski-runs for skiing. Thus, the penetration of surface water into the ground accelerated the creep of the slopes and finally triggered the landslide.

3.2 Landslide Analysis

Landslide analyses were done considering both the original ground conditions and the new conditions aroused from the cuts and fill to form the standard ski-runs of the complex. Owing to the less-consolidated and highly fractured nature of Gelinkaya formation, it was difficult to take undisturbed soil specimens during boring. However, those managed to be taken, wherever possible, were subjected to relevant laboratory tests to determine the geotechnical parameters essential to slope stability analyses. The total unit weight, cohesion intercept, and the internal friction angle of the representative samples varied from 19.1 to 20.2 kN/m^3, 15 to 30 kPa, and 27 to 30° respectively. The direct shear test was used to determine the shear strength parameters of the specimens under drained condition since no groundwater was encountered up to 100 m of boring depth. In addition, surface water which penetrated into the ground drained easily due to the porous and fractured nature of local soil. Therefore, the geotechnical parameters used in slope stability analysis were effective ones. Furthermore, the total unit weight, cohesion intercept, and the internal friction angle of the earthfill material placed at the top of the in-runs were found to be 20 kN/m^3, 40 kPa, and 29° respectively. Slope stability analyses regarding the most critical sections (K125 and K95 ski-runs) were performed by Slide software and their results are presented under the following sub-headings. The geotechnical parameters for each layer of slope profiles were given in Table 1 as well as in the legends of Figs. 9, 10 and 11.

Table 1. The geotechnical parameters used in slope stability analyses

Slope profile	Total unit weight (kN/m^3)	Cohesion intercept (kPa)	Internal friction angle (°)
K95	19	15	27
K125			
Lower layer	20	30	28
Upper layer	19	15	27
Earth fill	20	40	29

Pre-construction

The two most critical sections of the original ground condition of Erzurum ski-jumping complex area was obtained from the related topography map. They were where the K125 and K95 in-runs were constructed. In addition, the sections of slope profiles were constructed in reference with data procured from the site reconnaissance, in-situ, and laboratory test. The results of slope stability analyses by Slide software are given below in Fig. 8. The factor of safeties against the slope failure of the virgin ground under the static condition for both in-run areas were determined to be 1.424 and 1.484 respectively by modified Janbu method (Fig. 9).

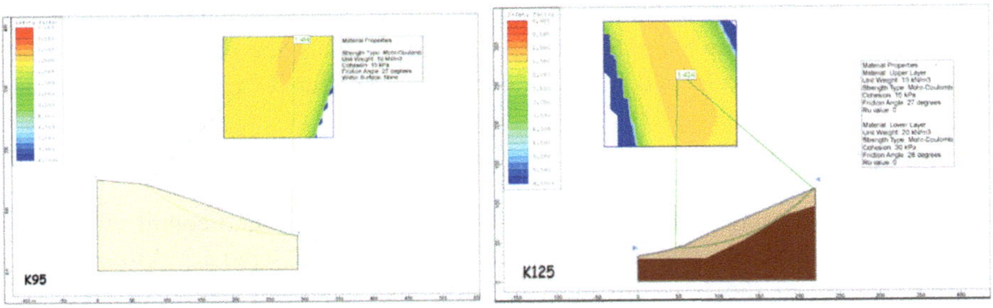

Fig. 9. Slope stability analyses of the two most critical sections

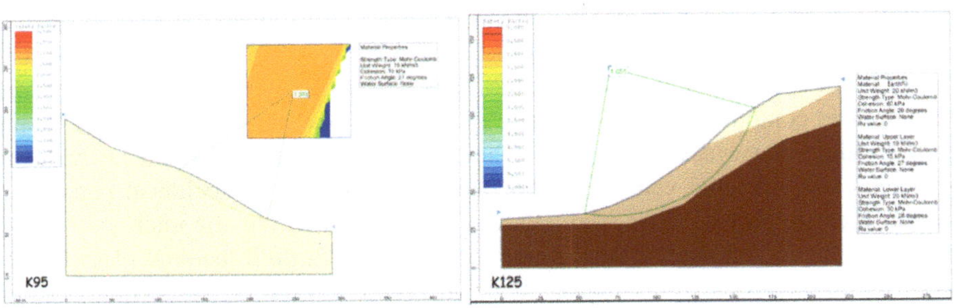

Fig. 10. Slope stability analyses of the reshaped sections

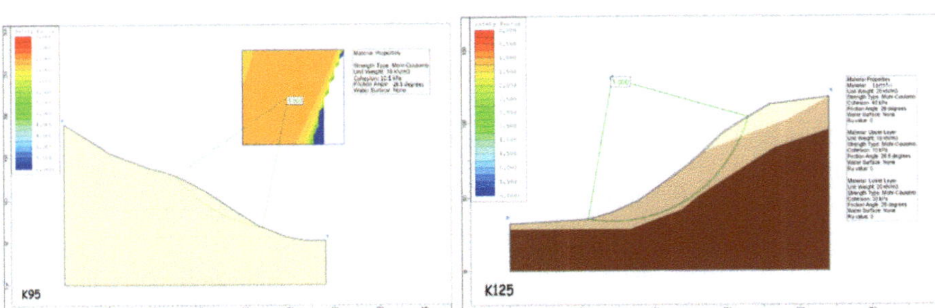

Fig. 11. Back analyses of the reshaped sections

As shown, the virgin ground slope of the two area under consideration had no instability problem at all.

Post-construction

To reshape the original slopes of the areas in question for the K125 and K95 ski-runs, the original slopes were steepened by cut-and-fill and their weights were increased by the surcharges due to earthfills. In fact, these processes brought the original slopes to a limit equilibrium condition under the static condition as analyzed by modified Janbu method (Fig. 10).

Back analysis has been proven to be one of a useful tool for the investigation of slope failures. It determines the shear strength parameters needed to generate the desired factor of safety. This analysis need the data such as geometry of slope debris, pore water pressure, density of failure material and the strength of material along the slip surface. There are several methods to do back analysis. However, in this case-study, the shear strength parameters, namely cohesion intercept and friction angle which resulted in a safety factor of 1 were searched by trial and error for K125 and K95 in-runs' slopes to compare between the input data for the back analysis and the situation observed in the field including laboratory data. For the rearranged slope conditions, the shear strength parameters determined in this way (Fig. 11 and Table 2) were compared with those used in the previous analyses (Fig. 10 and Table 1); they were found to be very close to each other.

Table 2. Comparison of shear strength parameters obtained from laboratory tests and back analyses for the soil profiles under consideration

Slope profile	Cohesion intercept (kPa)	Internal friction angle (°)	Data
K95	15	27	Laboratory
	10.1	26.5	Back analysis
K125 Upper layer	15	27	Laboratory
	10	26.6	Back analysis
Lower layer	30	28	Laboratory
	30	28	Back analysis

The above analyses revealed that the depth of failure surface was about 25 m and the slope failure included some part of the earthfill at the top of the slopes. All these were compatible with the actual situation encountered in the field.

4 Conclusions

1. Geological, geophysical and geotechnical data and their evaluation were quite insufficient.
2. Slope stability analyses of the original ground and the rearranged conditions formed by cuts and fills were not taken into account.
3. Project planning lacked in the effective drainage system, especially including the five ski-runs' area.
4. This case story could illustrate how important and cost-effective are comprehensive geotechnical engineering site investigation before the construction of infrastructure in risk areas.
5. To avoid geotechnical risk and economic losses localization of this type of sport infrastructure could be carefully planned in advance and included detailed recognition of geotechnical engineering condition.

6. Same of the areas could not be suitable for this type of infrastructure at all or cost of remediation works could be too high.

References

Petley, D.: Global patterns of loss of life from landslides. Geology **40**, 927–930 (2012)

Sorbino, G., Nicotera, M.V.: Unsaturated soil mechanics in rainfall-induced flow landslides. Eng. Geol. **165**, 105–132 (2013)

Kim, J., Jeong, S., Park, S., Sharma, J.: Influence of rainfall-induced wetting on the stability of slopes in weathered soils. Eng. Geol. **75**, 251–262 (2004)

Fredlund, D.G., Rahardjo, H., Fredlund, M.D.: Unsaturated Soil Mechanics in Engineering Practice. Wiley, New York, USA (2012)

Lu, N., Godt, J.: Infinite slope stability under steady unsaturated seepage conditions. Water Resource. Res. **44**, 11 (2008)

Sidle, R.C., Pearce, A.J., O'Loughlin, C.L.: Hillslope stability and land use. Water Resources Monograph. vol. 11. The American Geophysical Union. Washington, DC. 11, 140 (1985)

Rahardjo, H., Leong, E.C., Gasmo, J.M., Tang, S.K.: Assessment of rainfall effects on the stability of residual soil slopes. In Proceedings of the 2nd International Conference on Unsaturated Soils, vol. 2, pp. 280–285. Beijing, China (1988)

Ng, C.W.W., Zhan, L.T., Bao, C.G., Fredlund, D.G., Gong, B.W.: Performance of an unsaturated expansive soil slope subjected to artificial rainfall infiltration. Géotechnique **53**, 143–157 (2003)

Ng, C.W.W., Springman, S.M., Alonso, E.E.: Monitoring the performance of unsaturated soil slopes. Geotech. Geol. Eng. **26**, 799–816 (2008)

Yılmaz, A., Uysal, Ş.: 1:100 000 scale- geological mapping of Turkey – Map section F32. MTA publication. Ankara. 20 p. (1988)

Geological Investigation and Risk Assessment for Disaster Management of Merapi Volcano and Surrounding Area, Yogyakarta Special Territory, Indonesia

Ayu Narwastu Ciptahening[1(✉)], Nandra Eko Nugroho[1], and Noppadol Phienwej[2]

[1] Universitas Pembangunan Nasional "Veteran" Yogyakarta, Yogyakarta, Indonesia
ciptaheningayu@gmail.com
[2] Asian Institute of Technology, Khlong Luang, Thailand

Abstract. Mount Merapi is classified as one of the most attractive volcanos in the world. It is situated in Central Java and Yogyakarta Special Territory, Indonesia. It regularly erupts in every 4 years. Deposits produced by the activity of Mount Merapi are spread out in Sleman Regency Yogyakarta Special Territory, where the study area is located. The area includes four rock units, specifically Ancient Merapi, Middle Merapi, Recent Merapi, and Modern Merapi.

Mount Merapi eruptions had been responsible of the collapse of Hindu Mataram Kingdom, Central Java, in the 10th to 11th century periods. This was demonstrated by the discovery of many ancient temples in Yogyakarta and surrounding areas, such as Temple Sambisari, Temple Kadisoka, and Temple Kedulan which were buried by volcanic deposits of Mount Merapi. The eruption of Mount Merapi, in the year 2010, recognized as the biggest eruption in recent decades, claimed about 300 lives and very large losses. The type of eruption changed from Merapi type in the activity period of 2006 and before to become the combination of Vulcanian type and Pelean type. Due to the very intensive activities of the volcano, the geohazard threats faced by people living in the surrounding area of Yogyakarta are high. And in the face of the volcanic disasters, the people of Mount Merapi unite in the implementation of disaster preparedness.

In the study area, the geohazards are represented by lava flow, pyroclastic flow, ash fall, and lahar. There are four areas can reached by pyroclastic flow, including the Turi, Pakem, Cangkringan, and Ngemplak districts. All of the districts in the study area are threatened by ash fall, such as the Turi, Pakem, Cangkringan, Tempel, Sleman, Ngaglik, Ngemplak, Minggir, Sayegan, Mlati, Gamping, Godean, Depok, and Kalasan districts. While the areas threatened by lahar are Turi, Pakem, Cangkringan, Tempel, Ngaglik, Ngemplak, and Kalasan districts. By applying a scoring method, the study area can be classified into zone with high, moderate, and low geohazard vulnerability, threat, and risk.

Keywords: Volcano · Mount Merapi · Geohazards · Vulnerability Threat · Risk

J. Wasowski and T. Dijkstra (Eds.): GeoMEast 2018, SUCI, pp. 49–59, 2019.
https://doi.org/10.1007/978-3-030-02032-3_5

1 Introduction

1.1 Background

Indonesia is rich in natural resources and culture, but also suffers from many natural disasters. The presence of geohazards is related to the tectonic setting of Indonesia located at the juncture of three big lithosphere plates, i.e. Indo-Australia plate in the south, Eurasia plate in the north, and Pacific plate in the east. Thus Indonesia belongs to a seismically and volcanically active region known as the "ring of fire". Indeed, Indonesia has many active volcanoes.

One of the most active volcanoes in Indonesia is Mount Merapi located at the northern part of Yogyakarta Special Territory. The eruption of Mount Merapi in the year 2010 had made headlines around the world either through mass media or print media. This was the biggest eruption in recent decades and it claimed lives and very large losses. In the historical records, the most dangerous and destructive threats of Mount Merapi activities are pyroclastic flows and debris/mud flows (lahars). There are many changes that occur on geological aspects, benefits, and losses caused by the eruption of Mount Merapi 2010. Due to the existence of Mount Merapi that is very active, it can be underlined that the geohazards faced by people living in the surrounding area of Yogyakarta is about the impact of volcanic eruptions.

Based on the above backgrounds, it is necessary to conduct Geological Investigation, Risk Assessment and Cultural Behaviour Analysis for Disaster Management of Merapi Volcano and Surrounding Area.

1.2 Objectives

The objectives of geological investigation in the Mount Merapi and surrounding area are to study on geology and risk assessment for identifying the threat types and mapping the geohazard distribution in the study area and performing a risk assessment to determine the anticipation and prevention efforts in order to develop disaster management.

1.3 Location and Accessibility of the Study Area

The study area is located in the Sleman regency, Yogyakarta Special Territory, Indonesia. It is bordered by Magelang regency in the West, Boyolali regency in the North, Klaten regency in the East, and Bantul regency in the south, with the total area of 32 km x 32 km. Geographically, the study area lies between latitudes 7° 33' 05.8"–7° 46' 41.0" South latitude and longitude 110° 16' 16.2"–110° 29' 18.4" East longitude, and astronomically, the study area is located in 419595 mE–443595 mE and 9140205mN–9165205mN WGS 1984 UTM (Universal Transverse Mercator) (Fig. 1).

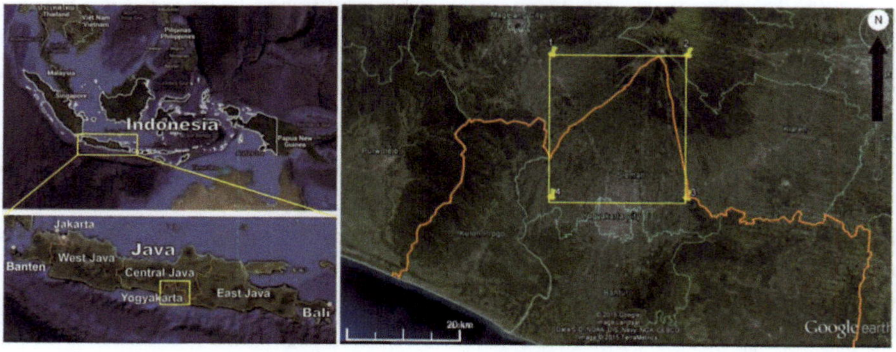

Fig. 1. Location of the study area

2 Method of Study

In this research carried out analytical and descriptive approaches, by processing data both primary and secondary. The research methods applied in this study includes:

a. Literature review is done by studying the existing references of previous researchers regarding the study area.
b. Collecting of secondary data includes the results of previous researches on the geomorphology, geology (stratigraphy and geological structures), and geohazards.
c. Interpretation and analysis of topographic map and remote sensing. The map used for the interpretation is a topographic map of 1: 75,000 scale, aerial photographs of 1: 30,000 scale, as well as satellite imagery.
d. Geological survey, mapping, and rock sampling. Field survey is done to check the validity of some secondary data, including geological conditions, the spreading of lithology and geological structure in the study area. On the other hand, geological mapping itself is implemented from field data collection and rock sampling. The field works are included conducting path, observation, measurements, sampling, data documentation, recording and plotting of data into the field map.
e. Laboratory testing and analyses represented Petrologic testing and thin section analyses are done to determine the type and the name of rocks.
f. Risk analysis and assessment is done as a basic of the preparation for risk reduction and anticipation to the geohazard threat that may occur.

3 Result and Discussion

3.1 Geology

3.1.1 Geomorphologic Units in the Study Area

Morphology of a volcano is the result of interaction between endogenous and exogenous processes. The morphology of a volcano is not only developed by the material released and the type of eruption, but also by the degree of volcanic activity, erosion, as well as the geologic structure. The study area is included in volcanic landform, and divided to four geomorphologic units, they are:

Parasitic Cone (V1)
This unit is developed by volcanic process. It occupies 5% of the entire study area. It is a volcanic hill with steep to very steep slope which is 21%–140% and has a range of elevation 1000–1400 m above sea level (asl). Lithology occupying Parasitic Cone unit are lava, autoclastic brecciated lava, and pyroclastic breccia. This unit is used for cultivation of tree planting area and for foresting.

Upper Volcanic Slope (V2)
This unit is formed and influenced by volcanic process, occupying 10% of the entire study area, with steep to very steep slope which is 21%–140% and has a range of elevation 1350–2600 m asl. Upper Volcanic Slope unit is occupied by lava flow deposits, pyroclastic breccia, laharic breccia, pyroclastic and laharic deposits. This unit is usually used for cultivation area and farming land.

Middle Volcanic Slope (V3)
This unit is also formed and influenced by volcanic process, occupying 15% of the entire study area, with steep slope which is 21%–55% and has a range of elevation 600–1350 m asl. Middle Volcanic Slope unit is occupied by pyroclastic breccia, laharic breccia, fine grained pyroclastic deposits as sandstone, laharic deposit of unconsolidated boulders, gravels, and sand. The unit is mostly used for agricultural area. There are some villages located in this area.

Lower Volcanic Slope (V4)
This unit is formed mostly by volcanic deposition. It occupies 70% of the entire study area, with sloping slope which is 8%–13% and has a range of elevation 150–600 m asl. Lithology occupying the unit are generally unconsolidated pyroclastic and laharic deposits. Lower Volcanic Slope unit is in general used for agricultural and residential areas. There are many villages and town located in this unit (Fig. 2).

3.1.2 Rock Units/Litology of the Study Area

According to Camus et al. (2000), there are four periods of Merapi volcanic activities, comprising Ancient Merapi period, Middle Merapi period, Recent Merapi period, and Modern Merapi period.

In general, lithology of the study area is dominated by volcanic deposits of Mount Merapi that can be grouped into the following rock units; Unit of Turgo Plawangan Lava Deposit, Unit of Ancient Merapi Pyroclastic Deposit, Unit of Ancient Merapi Laharic Deposit, Unit of Middle Merapi Lava Deposit, Unit of Middle Merapi

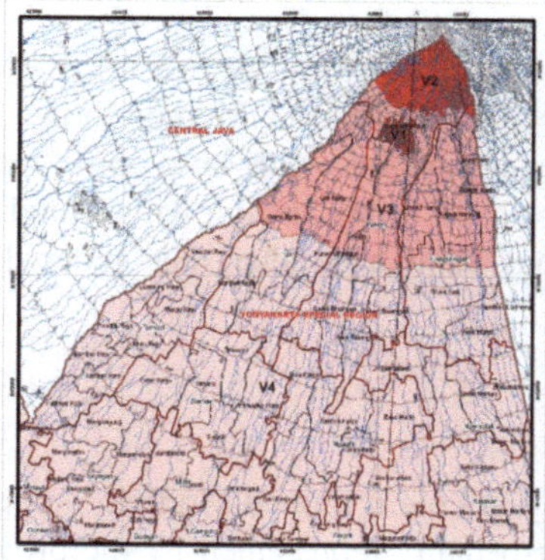

Geomophology aspect			Unit of Geomorphic: Parasitic cone (V1)	Upper volcanic slopes (V2)	Middle volcanic slopes (V3)	Lower volcanic slopes (V4)
Morphology	Morphography		Hill	Slope	Slope	Slope
	Morphometry	Slope	Very steep to steep	Very steep to steep	Steep	Sloping
		Area (%)	5	10	15	70
		Elevation (m)	1000-1400	1350-2600	600-1350	150-600
		Flow pattern	Radial	Radial	Radial	Radial
		Shape of valey	V	V	U	U
Morphogenesys	Morphostucture active		Volcanism	Volcanism	Volcanism	Volcanism
	Morphostucture passive		Lithology: pyroclastic breccia, laharic breccia pyroclastic and laharic deposit	Lithology: pyroclastic breccia, laharic breccia pyroclastic and laharic deposit	Lithology: pyroclastic breccia, laharic breccia pyroclastic and laharic deposit	Lithology: laharic and fluvial deposit
	Morphostructure dynamic		Erosion, Wethering	Erosion, Wethering	Erosion, Wethering	Erosion, Wethering

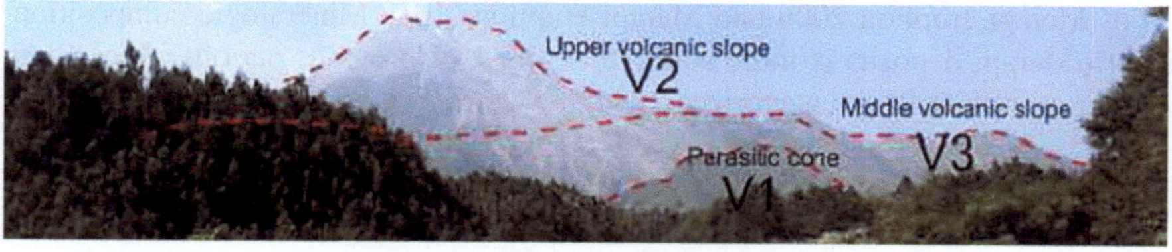

Fig. 2. Geomorphology map of the study area

Pyroclastic Deposit, Unit of Recent Merapi Lava Deposit, Unit of Recent Merapi Pyroclastic Deposit, Unit of Recent Merapi Laharic Deposit, Unit of Modern Merapi 2006 Pyroclastic Deposit, Unit of Modern Merapi 2006 Lahar Deposit, Unit of Modern Merapi 2010 Pyroclastic Deposit, Unit of Modern Merapi 2010 Lahar Deposit and Unit of Alluvial Deposit.

The study area is geologically controlled by volcanic activity. The constituent rocks are composed of Ancient Merapi deposit, Middle Merapi deposit, Recent Merapi deposit, and Modern Merapi deposit.

Ancient Merapi Deposit

Ancient Merapi period is 14,000 to 40,000 years old. It produced lava, auto breccias, pyroclastic and laharic breccias. The composition of the deposits is basaltic andesite.

Middle Merapi Deposit

Middle Merapi period produced thick lava deposit and pyroclastic breccias. The lava can be obtained at Batulawang and Gajahmungkur hills, with andesitic composition. Age of Middle Merapi period is 2,200 to 14,000 years. In the study area, the deposit is represented by pyroclastic breccias, found in the northern part of Pakem and Cangkringan districts.

Recent Merapi Deposit
Recent Merapi period produced deposits consists of thin lava (less than 3 m), pyroclastic flow and laha deposits of 1,000 to 2,000 years age. Recent Merapi shows a particular character of activity called Merapi Type. It specifies with lava dome formation, followed by lava avalanches forming pyroclastic flows and glowing clouds. The deposit consists of lava, pyroclastic, tuffaceous sand and lahar that generally unlithified yet, with andesitic composition.

Modern Merapi Deposit
The product of Mount Merapi that is less than 1000 years old is named Modern Merapi Deposit. In this study, Modern Merapi Deposit is especially represented by the products of Merapi eruption 2006 and Merapi eruption 2010. Mineralogic composition of Modern Merapi deposits is more acidic than those of previous periods. According to Hammer et al. (2000), it categorized as high *K-basaltic andesite* (Fig. 3).

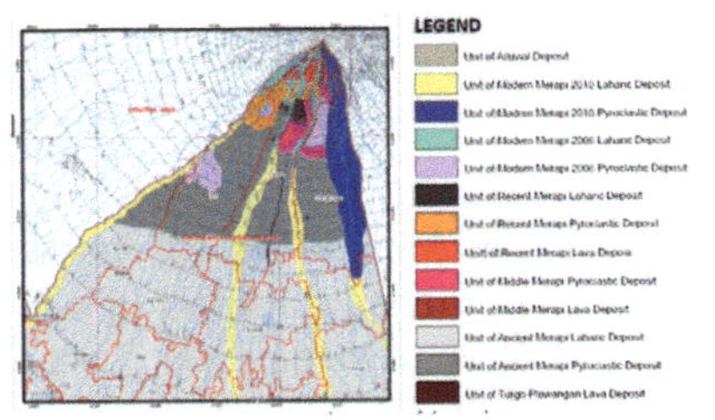

Fig. 3. Geologic map of the study area

Merapi 2006
Physically, the deposits of Merapi 2006 were resulted both from primary and secondary activities, composed of pyroclastic flow, surge, and lava, with grain size of bolder, sand, ash, and dust

Merapi 2010
Mount Merapi 2010 produced pyroclastic flow which is deposited more than 15 km distant from the center of the eruption to the southeast direction through River Gendol valley. There were 100 million m^3 volcanic materials deposited on the upper slopes of the volcano.

Alluvial Deposits
Alluvial deposits are especially composed of volcanic materials that weathered, eroded, transported through river and finally deposited. This unit is found in the southern part of the study area, in form of floodplains and bars.

3.2 Geohazards in the Study Area

In the study area, geohazards are related to volcanic activities, therefore represent volcanic hazards including lava flow, pyroclastic flow, ash fall, and lahar.

3.2.1 Lava Flow

The lava of Mount Merapi is relatively viscous (Kusumayudha 2013), the temperature varies from 800 °C to 1000 °C. Due to its viscosity, lava flow of Merapi doesn't move fast. Its flow distance reaches maximally 1,5 to 4,5 km from the eruption center, thus people very rare killed by it, even though it is extremely hot (800 °C–1000 °C). The petrologic composition of Merapi lava is andesitic to basaltic andesite. By interpretation on topographic map which is delineation from contours patterns, in the study area, there are three districts that are threatened by lava flow. Those are the Turi, Pakem, and Cangkringan distict. According to the records of Goelogical Survey of Indonesia, the farthest of lava flow reached 4.5 km in the Pakem district.

3.2.2 Pyroclastic Flow

The most hazardous of Mount Merapi volcanic hazards is pyroclastic flow or glowing cloud. Pyroclastic cloud with temperatures between 600 °C–800 °C is able to burn all living creatures and various objects. The velocity is very high, up to 120 km/h resulted in any difficulty to escape from its attack. The composition of pyroclastic flow is tuff (ash), lapilli, volcanic sands, bombs, and blocks; is able to reach more than 15 km distant from the eruption center. By interpretation using satellite image and field work, there are four districts that able to be reached by pyroclastic flow, including the Turi, Pakem, Cangkringan, and Ngemplak district (Fig. 4).

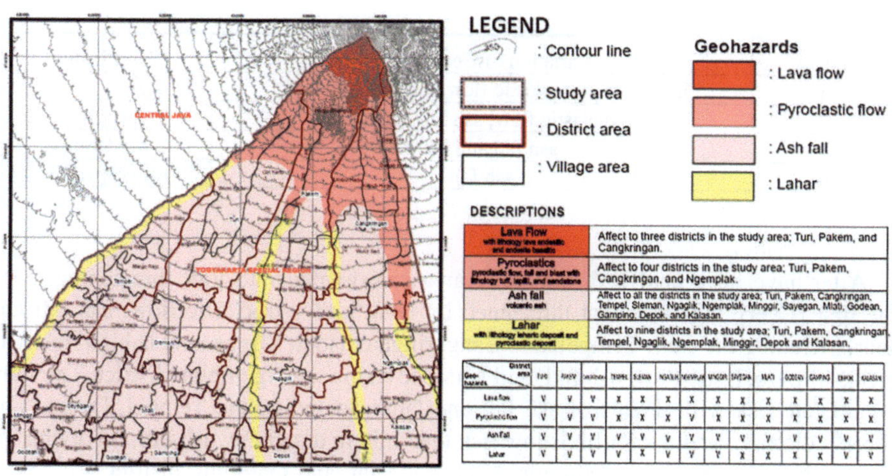

Fig. 4. Volcanic hazard map of the study area

3.2.3 Ash Fall

Ash fall belongs to an important threat of volcanic eruptions on the surrounding environment, because it can cause respiratory disease to humans, damaging crops, disrupting traffic in urban traffic including air flights, damaging infrastructure and vehicles. The ash of Merapi contains volcanic tuff and fine sands, with mineral composition of dominantly plagioclase, amphibole, and silica (SiO2). During eruption, ash fall can cover all districts in the study area, in another word, all of the study area is threatened by this hazards, such as the Turi, Pakem, Cangkringan, Tempel, Sleman, Ngaglik, Ngemplak, Minggir, Sayegan, Mlati, Gamping, Godean, Depok, and Kalasan district.

3.2.4 Lahar

In general, it occurs in rainy season. Lahar is composed of water and materials with various grain size from sand to very big boulder. The density of lahar is very high, flowing lead by gravitational force; the speed of flowing reaches 60 km/h. The districts affected by lahar are the Turi, Pakem, Cangkringan, Tempel, Ngaglik, Ngemplak, Mingggir, Depok, and Kalasan district.

3.3 Volcanic Hazard Threatening Level

As it has been mentioned above, there are four kinds of geohazards exist in the study area, including lava flow, glowing clouds or pyroclastic flow, ash fall, and mud flow or lahar. Qualitatively the volcanic hazard threatening level can be written as follows (Table 1):

Table 1. Volcanic hazard threatening level

Level	Variation of Geohazards	Classification
1	lava flow, pyroclastic flow, ash fall, and lahar	Very High
2	pyroclastic flow, ash fall, and lahar	High
3	ash fall and lahar	Moderate
4	ash fall	Low

After have been analyzed, the study area can divided into four threatening level. The first is very high risk that threatened by lava flow, pyroclastic flow, ash fall and lahar, including the Turi, Tempel, and Cangkringan distict; The second is high risk with threats of pyroclastic flow, ash fall, and lahar, covering the Ngemplak district; The third is moderate risk with threats of ash fall and lahar, including the Tempel, Depok, and Kalasan dictrict; And the fourth is low risk that only in danger by ash fall, involving the Sleman, Ngaglik, Sayegan, Minggir, Mlati, Godean, and Gamping district (Fig. 5).

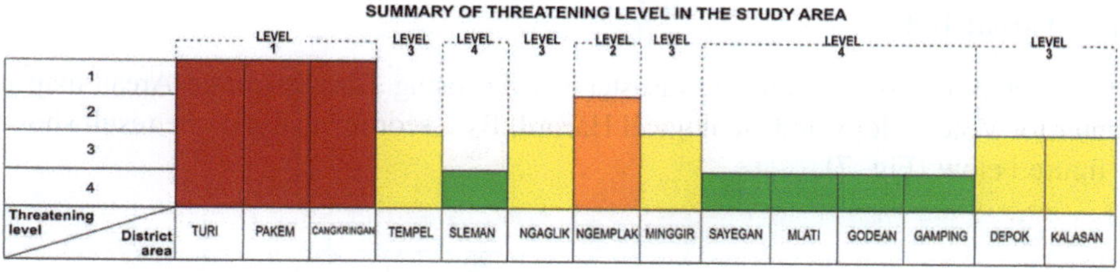

Fig. 5. Summary of volcanic hazards threatening level

3.4 Vulnerability of the Study Area

Vulnerability is situations or conditions which are cause inability of the people to prepare themselves to face natural disaster. Vulnerability covers very diverse. The general situations that include factors of physical, socio-economy, political, and cultural, which are could potentially give effect to the people to more easily affected by disasters. The vulnerability in the study area can be classified in three vulnerability classification, those are high vulnerability, moderate vulnerability, and low vulnerability. To classify vulnerability in the study area, there are four main factors looked up; Human, Infra-structure, Agriculture, and Socio ecomony (cost losses in Rupiah).

3.5 Risk Assessment

Disaster risk assessment is an approach to show potential negative impacts that may arise as a result of a potential disaster struck. Potential negative impacts are calculated based on the level of vulnerability and capacity of the region. The potential negative impact is seen from the potential number of people who are exposed to, loss property and environmental damage (Fig. 6).

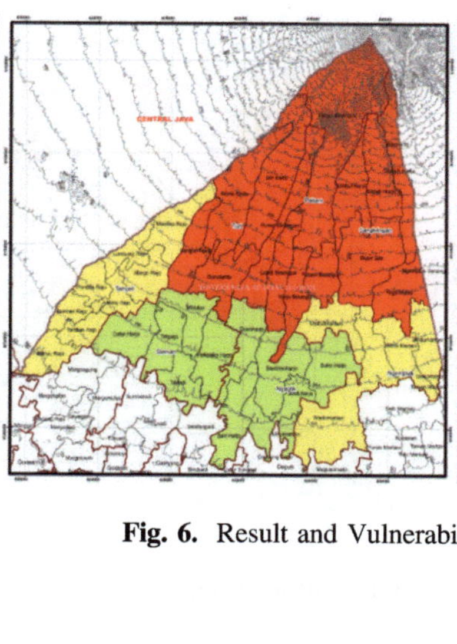

NO	Village	Description
1	TURI	HIGH
2	PAKEM	HIGH
3	CANGKRINGAN	HIGH
4	TEMPEL	MODERATE
5	SLEMAN	LOW
6	NGAGLIK	LOW
7	NGEMPLAK	MODERATE

Fig. 6. Result and Vulnerability Map of the study area

3.6 Threat Index

Threat determination of volcanic disaster viewed using Disaster-prone Areas map of Center for Volcanology and Geological Hazard. By a scoring analysis, the result shown in figure below (Fig. 7).

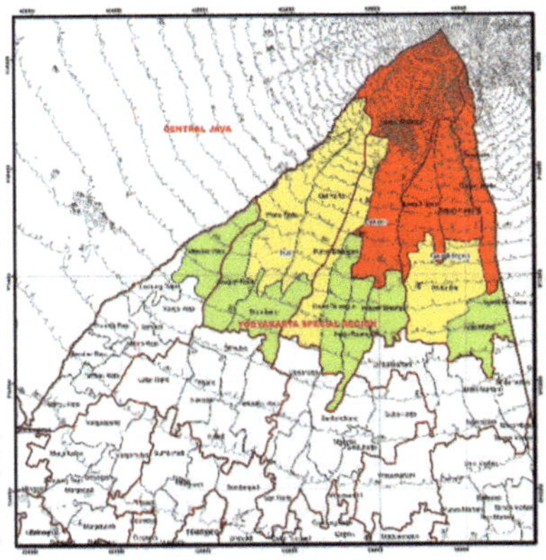

TURI		PAKEM		CANGKRINGAN	
Village	Description	Village	Description	Village	Description
Giri Kerto	MODERATE	Hargo Binangun	HIGH	Glagah Harjo	HIGH
Wono Kerto	MODERATE	Purwo Binangun	MODERATE	Umbul Harjo	HIGH
Bangun Kerto	LOW	Candi Binangun	LOW	Kepuh Harjo	HIGH
Dono Kerto	LOW	Pakem Binangun	LOW	Wukir Sari	MODERATE
		Harjo Binangun	LOW	Argo Mulyo	LOW

TEMPEL		SLEMAN		NGAGLIK		NGEMPLAK	
Village	Description	Village	Description	Village	Description	Village	Description
Mardiko Rejo	LOW	Trimulyo	SECONDARY	Suko Harjo	SECONDARY	Umbulmartani	SECONDARY
Lumbung Rejo	SECONDARY	Pendowo Harjo	SECONDARY	Donoharjo	SECONDARY	Bimo Martani	SECONDARY
Pondok Rejo	SECONDARY	Tridadi	SECONDARY	Sardonoharjo	SECONDARY	Wedomartani	SECONDARY
Margo Rejo	SECONDARY	Triharjo	SECONDARY	Sinduharjo	SECONDARY	Sindumartani	SECONDARY
Moro Rejo	SECONDARY	Catur Harjo	SECONDARY	Sari Harjo	SECONDARY		
Sumber Rejo	SECONDARY			Minomartani	SECONDARY		
Tambak Rejo	SECONDARY						
Banyu Rejo	SECONDARY						

Fig. 7. Result and Threat Map of the study area

3.7 Disaster Risk Level

Disaster Risk level is determined by combining the Level Capacity Loss Rate. Disaster Risk Level Determination conducted for each potential hazard in an area. By a scoring analysis, the result shown in the figure below (Fig. 8).

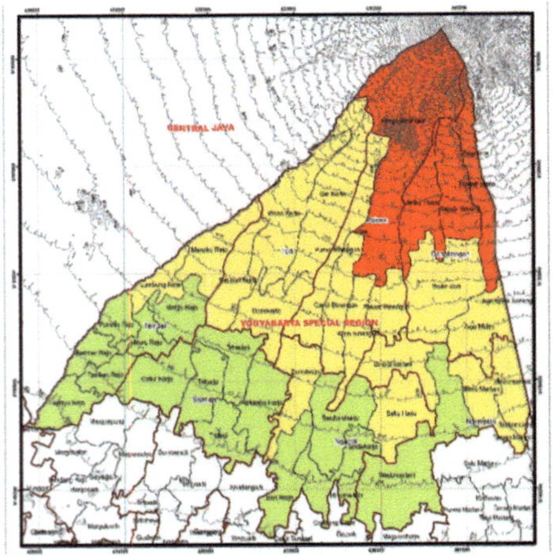

TURI		PAKEM		CANGKRINGAN	
Village	Description	Village	Description	Village	Description
Giri Kerto	MODERATE	Hargo Binangun	HIGH	Glagah Harjo	HIGH
Wono Kerto	MODERATE	Purwo Binangun	MODERATE	Umbul Harjo	HIGH
Bangun Kerto	MODERATE	Candi Binangun	MODERATE	Kepuh Harjo	HIGH
Dono Kerto	MODERATE	Pakem Binangun	MODERATE	Wukir Sari	MODERATE
		Harjo Binangun	MODERATE	Argo Mulyo	MODERATE

TEMPEL		SLEMAN		NGAGLIK		NGEMPLAK	
Village	Description	Village	Description	Village	Description	Village	Description
Mardiko Rejo	MODERATE	Trimulyo	LOW	Suko Harjo	MODERATE	Umbulmartani	MODERATE
Lumbung Rejo	MODERATE	Pendowo Harjo	LOW	Donoharjo	MODERATE	Bimo Martani	MODERATE
Pondok Rejo	LOW	Tridadi	LOW	Sardonoharjo	LOW	Wedomartani	LOW
Margo Rejo	LOW	Triharjo	LOW	Sinduharjo	LOW	Sindumartani	MODERATE
Moro Rejo	LOW	Catur Harjo	LOW	Sari Harjo	LOW		
Sumber Rejo	LOW			Minomartani	LOW		
Tambak Rejo	LOW						
Banyu Rejo	LOW						

Fig. 8. Results and disaster risk map of the study area

4 Conclusions

Based on the research and analysis that has been done in research, as well as the discussion in the previous chapters, it can be concluded as follows:

1. Geohazards are natural occurrence that involves geological processes. In the study area, geohazards that occur are related to volcanic activities, therefore it can be named volcanic hazards that consist of lava flow, pyroclastic flow, ash fall, and lahar. There are three districts that are threatened by lava flow, those are the Turi, Pakem, and Cangkringan distict. There are four districts that able to be reached by pyroclastic flow, including the Turi, Pakem, Cangkringan, and Ngemplak district. All of the study area is threatened by ash fall, such as the Turi, Pakem, Cangkringan, Tempel, Sleman, Ngaglik, Ngemplak, Minggir, Sayegan, Mlati, Gamping, Godean, Depok, and Kalasan district. While areas which threatened by lahar are the Turi, Pakem, Cangkringan, Tempel, Ngaglik, Ngemplak, and Kalasan district.
2. By applying scoring method on geohazard vulnerability, the study area can be classified into three vulnerability classification, those are high vulnerability, moderate vulnerability, and low vulnerability.
3. Based on volcanic hazard threatening analysis have been done, the study area can divided into three volcanic hazard threatening classification. The first is high that including Hargo Binangun village in the Pakem district and Glagah Harjo, Umbul Harjo, Kepuh Harjo village in Ngemplak district. The second is moderate that covering the Giri Kerto, Wono Kerto village in the Turi district, Purwobiangin village in Pakem district and Wukirsari village in Cangkringan district. The third is low that including the Mardirejo village in Tempel district, Binangun Kerto and Dono Kerto village in Turi district, Candi Binangun, Pakem Binangun, Harjo Binangun in Pakem district, and Argomulyo village in Cangkringan district.
4. Accordung to risk analysis, the study area included in high, moderate, and low risk area. The high risk area covered Hargo Binangun village of the Turi district, and Glagah Harjo, Umbul Harjo, Kepuh Harjo village of Ngemplak district. The moderate risk area located in the Mardiko Rejo, Lumbung Rejo village of Tempel district, Giri Kerto, Wono Kerto, Bangun Kerto, Dono Kerto village of Turi district, Purwo Binangun, Candi Binangun, Pakem Binangun, Harjo Binangun village of Pakem district, Wukir Sari, Argo Mulyo village of Cangkringan district, Suko Harjo, Donoharjo village of Ngaglik district, and Umbulmartani, Bimo Martani, Sindumartani village of Ngemplak district.

References

Kusumayudha, S.B.: Gunungapi Aktif di Indonesia (Active volcanoes of Indonesia), 2nd edn. PT Citra Aji Parama, Yogyakarta (2013)

Toyo Santoso Dipo, H.: Bencana Tanah Longsor Kabupaten Kulon Progo dan Upaya Mitigasi Bencana, Pros. Simposium Nasional Pencegahan Bencana Sedimen, pp. 40–49 (2002)

Van Bemmelen, R.W.: The Geology of Indonesia, vol. IA, 732 pp. Martinus Nijhoff, The Haque (1949)

Acquisition and Analysis of Surface Wave Data in the Indo Gangetic Basin

P. Anbazhagan[1,2(✉)], Ketan Bajaj[1], Sayed S. R. Moustafa[2,3], and Nassir S. N. Al-Arifi[2]

[1] Department of Civil Engineering, Indian Institute of Science, Bangalore, India
anbazhagan@iisc.ac.in

[2] Geology and Geophysics Department, Faculty of Science, King Saud University, Riyadh 11451, Saudi Arabia

[3] Seismology Department, National Research Institute of Astronomy and Geophysics (NRIAG), Cairo 11421, Egypt

Abstract. Seismic activity in the tectonically active Himalayan region can result in devastation in the contagious deep Indo-Gangetic Basin (IGB). It is filled by alluvial deposits and, therefore, prone to site amplification. IGB had experienced catastrophic earthquake damages due to the presence of thick soil depth of 0.05 km to about 4 km. However, very few studies have presented seismic site characterization of the IGB soil up to the shallow depth, and very limited attempts have been made to measure the dynamic properties of the deep soil column. Hence, in this study, shear velocity profile ($\mathbf{V_s}$) up to 400 m depth is measured using combined active and passive multichannel analysis of surface wave (MASW) survey in 275 selected locations in IGB. Acquisition and processing of the seismic surface wave data to obtain the $\mathbf{V_s}$ profiles at shallow and deep depths are discussed in this paper. Parametric study has been done to identify the variation in dispersion and $\mathbf{V_s}$ profiles by recording time, different sampling interval, different polarity and gain. It has been seen that changing any parameter during data recording may result in variation in $\mathbf{V_s}$. Further, the surveyed sites are classified and characterized based on time-averaged $\mathbf{V_s}$ in the upper 30 m depth as per National Earthquake Hazards Reduction Program seismic site classification. $\mathbf{V_{s30}}$ values vary from 157 to 1152 m/s in the entire stretch of IGB. This is the first time such extensive study was carried out in IGB for $\mathbf{V_s}$ determination at deep and shallow depths.

Keywords: Indo-Gangetic Basin · MASW · Data acquisition
Seismic site characterization · Passive data recording

1 Introduction

Local site conditions can have great influence on ground surface motion and structural damage by earthquakes. Two classic examples that emphasize the influence of site amplification due to local site effect are the 1985 Mexico and the 1989 Loma Prieta earthquakes. Similarly, many earthquakes in India (1934 Bihar-Nepal; 2001, Bhuj; 2015 Nepal earthquake) have also illustrated the importance of local site effects. The Indian subcontinent is one of the most seismically active regions in the world.

J. Wasowski and T. Dijkstra (Eds.): GeoMEast 2018, SUCI, pp. 60–72, 2019.
https://doi.org/10.1007/978-3-030-02032-3_6

The large strains are building along and within the plate boundary due to the ongoing collision between the Indian and Eurasian Plates. Moreover, crustal shortening increases the earthquake hazard, particularly in the northern part of the Indian Subcontinent. Various authors (Bilham and Ambraseys 2005; Bilham et al. 2001) have studied the seismotectonics of the Himalayan region and predicted the high seismicity along the entire stretch of the Main Boundary Thrust (MBT), the Main Central Thrust (MCT) and Indus-Tsangpo Suture (ITS). In the last two centuries, the Himalayan region has experienced many events of magnitude more than 7. The highly fertile and deep basin of Indo-Gangetic Basin (IGB), bound on the north side by the Himalayas, is one of the most populous areas. It is about 1000 km long to the south and filled-up by loose soil deposits. High seismicity region surrounds the IGB and this makes the damage scenario more destructive. Significant causalities to human life or infrastructure loss can result from any large earthquake in the future. Hence, there is a need to study the spatial variation of shear wave velocity in the IGB for determining the local site effect due to the presence of deep deposits contagious with high seismicity region. The proper characterization of deep soil deposits in the IGB also sets the priority for the determination of seismic site classification and the amplification factor for different periods due to local site effect.

Various authors (Boominathan et al. 2008; Anbazhagan and Sitharam 2008; Anbazhagan et al. 2010; Kumar et al. 2012; Desai and Choudhury 2014, 2015; Kumar et al. 2016) in India have attempted to determine the shear wave velocity at shallow depths in different study areas. Anbazhagan and Sitharam (2008), determined the shear wave velocity profile for shallow depth in Bengaluru. Desai and Choudhury (2015) and Kumar et al. (2012) respectively derived the V_s profile for Mumbai and Lucknow region in India. Jishnu et al. (2013) derived the V_s profiles at four different locations for Kanpur for site response study. Pandey et al. (2016) derived the V_s profile of seismic stations in Delhi region by carrying the multichannel analysis of seismic waves. Most of the previous studies were limited to soil column of 30 m depth. Additionally, in many of the studies, the measured SPT-N values were converted to V_s profiles and used for site response studies. Till today there are no studies available for determining the local site effect for the deep deposits of IGB from the measured V_s profiles exceeding 100 m depth.

There are several approaches for estimating V_s through surface wave analysis and with the developments in geophysical-indirect methods, the degree of uncertainty is reducing eventually. These geophysical methodologies integrate border soil volume and, therefore, are more representative of the seismic behavior of a site than local in-hole measurements (Humire et al. 2015). In this study, combined active and passive multichannel analysis of surface wave (MASW) survey has been used for determining the shear wave velocity profile at 275 locations in the IGB. For evaluating the dispersion characteristic both active and passive sources, linear and circular arrays have been used. Geophones of 2 Hz frequency have been used for performing the MASW survey. Ambient noise has been used as a source for both passive remote and roadside survey. Recordings have been done at different sampling intervals and record lengths. Measurements are performed at different locations in the entire stretch of IGB, stretching across the Indian states from Bihar to Punjab. The sites are classified and characterized based on time-averaged V_s in the upper 30 m depth as per NEHRP (BSSC 2003) seismic site classification.

2 Study Area

The IGB is shaped as a result of post-collision between Indian and Asian tectonic plates (Singh 1996). It is a peripheral foreland basin system, formed on a flexed Indian plate lithosphere (Lyon-Caen and Molnar 1985; Singh 1987). The Ganga Plain spreads from Aravalli-Delhi Ridge to Rajmahal hills along the west and the east respectively; Himalayan foothills in the north to the Bundelkhand Vindhyan Plateau in the south, occupying an approximate area of 250,000 km^2, lies between 77°E to 88°E (longitude) and 24°N and 30°N (latitude). The length of IGB is about 1000 km; with the variable width, ranging from 200 to 450 km, being wider and narrower, along the western and the eastern part (See Fig. 1). The IGB is irregular along with outcrops of rocks pounding out of the alluvial in the southern edge, whereas, the Siwalik rocks are exposed in the northern margin beside the Himalayan Frontal Fault (HFF) (See Fig. 1). The IGB shows all the foremost components of a foreland basin system, namely an orogen (the Himalaya), deformed foreland basin deposits adjacent to the orogen (Siwalik Hills), a depositional basin (Ganga Plain) and peripheral cratonic bulge (Bundelkhand Plateau) (Singh 1996). IGB established in the Early Miocene, and from Middle Miocene to Middle Pleistocene, the northern part of the IGB was uplifted, and thrust basin-wards; the Ganga plains moved southwards because of thrust loading in the orogen (Singh 1996).

Based on the geophysical study and deep drilling carried out by ONGC, various researchers (Rao 1973; Lyon-Caen and Molnar 1985) depicted the basement structure and sub-surface geology. As per, Rao (1973), the basement thickness of alluvium is $\sim$6 km near the foothill zone and decreases gradually southwards. Addition to variable thickness of sedimentary fill, the Ganga Basin exhibits number of ridges, fills and depressions. The important basement highs are the Delhi-Hardwar ridge in the west, the Faridabad ridge in the middle, the Monghyr-Ghazipur ridge in the east, a poorly developed high in the Mirzapur-Ghazipur area and smaller "highs" of Raxaul, Bahraich and Puranpur. There are two important basins or low areas, namely Gandak and Sarda depressions (see Fig. 1). The densely populated alluvial plain is characterized by the presence of several transverse and oblique subsurface faults. A number of subsurface faults trending NW-SE, WNW-ESE and NE-SW, transverse to the trends of the Himalayan, were reported and most of these are known to be seismically active (Valdiya 1987). The IGB is spatially variable in terms of geomorphology, tectonics, differences in geological deposits, and dynamics of river deposition, all resulting in the presence of different lithological units. MASW survey has been carried out in the entire stretch of IGB to capture the spatial variability in shear wave velocity.

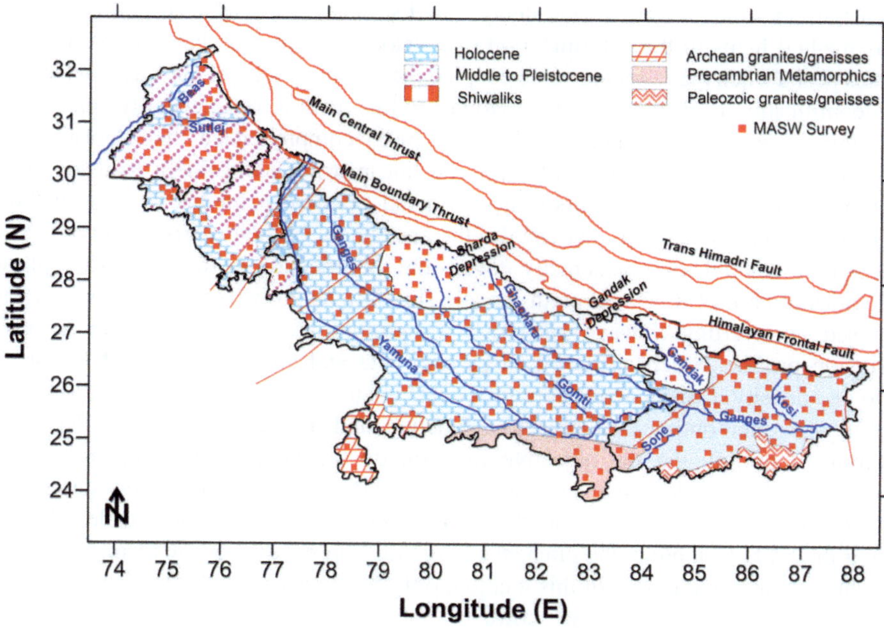

Fig. 1. Geological description of Indo Gangetic Basin along with the MASW survey locations

3 Field Survey

MASW is a geophysical survey that records Rayleigh waves on a multichannel record. It utilizes the dispersion properties of surface waves for determining the shear wave velocity profile in 1D and 2D format. Based on the source of the surface wave recorder, MASW is classified as Active or Passive MASW survey. In case of active MASW survey, surface waves are produced actively through an impact source like a sledgehammer. However, in case of Passive MASW survey, surface waves are produced passively by natural (e.g., thunder and tidal motion) or cultural (e.g., traffic) activities, (Park et al. 2007). The maximum reliable investigation depth is typically shallower than 50 m in case of active MASW survey. However, it can range a few hundred meters in case of passive MASW survey.

The entire process for developing V_s profiles through MASW involves three steps: procurement of ground roll, development of dispersion curve (a plot of phase velocity versus frequency), and back calculation (inversion) of V_s profile from the calculated dispersion curve. Multichannel methods permit a synchronized study of multiple geophone records which allow a transformation from time and space domains to another domain that acknowledges the recognition of energy peaks including the dispersive characteristics of the investigating site (Foti et al. 2001). Various approaches are available for this purpose. The most effective and widely used are the frequency-wavenumber analysis (f-k) and the spatial autocorrelation method (SPAC) proposed by Aki (1957). Analogous to f-k analysis, Park et al. (1998) recommended a transformation algorithm, in which all signals are delayed and summed for a selected phase velocity. This method helps in the determination of a velocity–frequency diagram that demonstrates peaks when the expected value coupled the phase velocity of the wave and allows the identification of dispersion curve. However, this method was generally appropriate for active measurements. Park and Miller (2008) comprehended this algorithm for ambient vibration measurement utilizing the linear arrays alongside a roadside. This method sums up the energy associated with all possible azimuths to construct the velocity–frequency diagram that enables the abstraction of the dispersion curve.

S-wave velocities fundamentally govern the modification in Rayleigh-wave phase velocities in case of a layered earth model. Thus, Rayleigh-wave phase velocities can be used in inverting S wave velocities. V_S profiles are determined from iterative inversion process which needs the dispersion data and approximations of Poisson's ratio and density. Least-squares approach helps in automation of the process (Xia et al. 1999), it further outlines an objective function and a weighting matrix, derives a solution of minimizing the objective function by the Levenberg–Marquardt (L-M) method and the singular value decomposition (SVD) technique. An iterative solution of a weighted damping equation using the L-M method provides a fast and stable solution. Using the SVD technique, the efficient calculation can be attained by reconstructing a weighted damping solution. In this method, only V_S is restructured after each iteration, with Poisson's ratio, density, however model thickness remaining unaffected throughout the inversion. As per Xia et al. (1999), the V_S profile is the most important parameter in the inversion process as fluctuations in density or Poisson ratio give insignificant effects in dispersion properties.

4 Data Acquisition and Processing

The raw data is acquired for 275 survey sites along the entire stretch of Indo Gangetic Basin. The data collection covers the major cities of Punjab, Haryana, Uttar Pradesh and Bihar states of India. Data acquisition sites are shown in Fig. 1. At each location, both Active and Passive MASW survey has been done to acquire the data at both shallow and deeper depths. Test setup consists of 24 channel Geode seismographs in combination with 24 vertical geophones with the frequency of 2.0 Hz. An impulsive source of 15-pound sledgehammer striking over a 30 cm × 30 cm size steel plate produces surface waves, in case of active survey. Figure 2 shows the acquisition of active data. However, the amount of energy generated by active source is not sufficient for producing dispersion curve at lower frequencies (e.g., 5–7 Hz). Hence to get the investigation depth of about several hundreds of meters, Passive noise is used. Passive surface waves generated from natural (e.g., tidal motion) or cultural (e.g., traffic) are used for generating low-frequencies (1–30 Hz). Depending on field conditions and type of V_S profiles to be obtained (1D or 2D); Passive MASW is divided into two types, i.e. passive remote and passive roadside MASW surveys. For acquiring the data for Passive MASW survey for the IGB both passive remote and roadside survey have been done; based on the availability of place (See Fig. 2). For obtaining the passive data, a passive roadside acquisition method is used by taking advantage of moving traffic for producing low-frequency ambient noise. Park and Miller (2008) recommended that when accomplishing a roadside surface wave using a linear receiver array, a 2-D dispersion analysis scheme that explains the offline nature of the passive surface waves is required. Hence, for obtaining the raw data using Passive survey, different sampling intervals and recording times are used to enhance the dispersion curve quality. After acquiring the data from both active and passive MASW survey, the individual dispersion curves have been extracted using velocity–frequency diagram.

Fig. 2. Multichannel analysis of surface wave instrumentation indicating the sources used in active and passive survey

The obtained raw data has been further processed to obtain the dispersion curve and finally to obtain the 1D shear wave velocity profile. Algorithm for obtaining dispersion curve and V_s profiles through inversion are explained above. For obtaining the V_s profiles, window-based software named 'SurfSeis 5' and 'ParkSEIS 2' has been used.

These software process the Rayleigh type seismic wave and generate V_s profiles by analyzing the fundamental mode of dispersion curve of Rayleigh wave. The dispersion curve for passive as well as active is given as Fig. 3(a) and (b) respectively. For (a) enlarging the analyzable frequency (therefore depth) range of dispersion and (b) identifying the modal nature of dispersion trends, it is useful to combine the dispersion images of active and passive data. Hence, combining the active and passive dispersion image has also studied to quantify the depth corresponding to both lower and upper-frequency range. For most of the data to get the enhanced shear wave velocity at lower as well as deeper depth combined dispersion is used. The dispersion curve for combined data is given as Fig. 3(c).

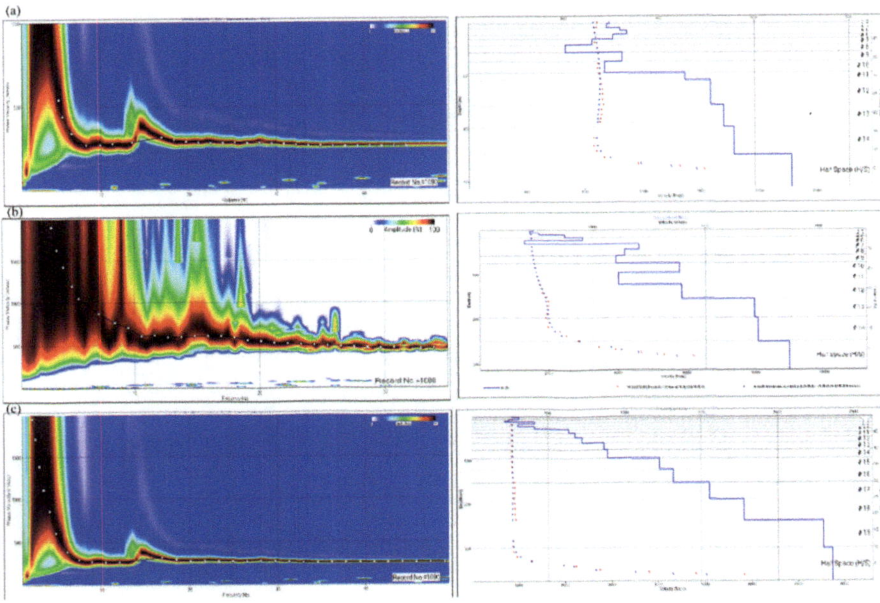

Fig. 3. Typical example of (a) Active, (b) Passive and (c) Combined MASW result for dispersion and shear wave velocity profile

5 Result and Discussion

5.1 Parametric Study on Dispersion Curve and vs Profiles

These are different ways of acquiring data using MASW. Different parameters, like recording time, sampling interval, number of geophones, geophone spacing governs the prediction of dispersion curve. This on other hand affects the layering and shear wave velocity of the site. Hence, in this study for different sites, these parameters are evaluated and variation on dispersion curve and V_s profiles are determined. For recording the passive data different recording time, sampling intervals, polarity and gain is tested. For example, data has been acquired for 120 s by changing the sampling interval as

8 ms, 4 ms and 2 ms and gain as high and low gain. Typical dispersion curve and V_s profiles for four cases have been given as Fig. 4(a). Case A1, A2, A3 and A4 corresponding to recording time interval of 60 s with sampling interval of 10, 8, 4 and 2 ms respectively. A significant change is observed at lower frequencies, due to that there is a variation in velocity at deeper depths. Standard deviation of 10 to 25 m/s is observed within a depth of 50 m. Further for the same site the data is recorded for different time intervals viz. 10, 30, 60 and 120 s. These cases are respectively referred as Case B1, B2, B3 and B4 in Fig. 4(b). Significant variation in dispersion curve is observed at lower frequencies, which affect the velocity at deeper depths. The difference of 150 to 200 m/s is observed at a depth more than 150 m. This may be due to the variability in traffic density and traffic loading. Hence, while capturing the passive data, recording must be done for different time intervals to obtain the proper energy bands at all the frequencies. Similarly, change in gain and for different recording time, there is a change in dispersion curve and corresponding variation in shear wave velocity. Hence for the final study, average of all the survey at same location has been used. You can also add how much error and difference.

Considering the dispersion curves and the corresponding velocity at different layers for various parameters for data recording following points are concluded

(1) The passive recording must be done for longer period, i.e. more than 60 s with larger sampling interval
(2) In case the traffic density is low, combined active and passive survey need to be done
(3) In case of poor energy bands, combining two passive images enhances the image to extract the dispersion curve
(4) To obtain the velocity at shallow as well as deeper depths both active and passive image need to be combined at the same site
(5) It is always good to do multiple experiments at a site, and average velocity need to be given.

5.2 Typical Comparison of vs with SPT-N Value and Lithology

SPT is a widely used in-situ test in a borehole to evaluate geotechnical properties of soil. The preliminary study has been done to compare the SPT N-value with the derived shear wave velocity at sites in Haryana, Punjab and Uttar Pradesh. All the recorded data has been used to compare the SPT N value and based on that input setting has been done for further data recording. The compared SPT N value and recorded V_s profile is given as Fig. 5(a). For different locations, V_s is compared with the SPT N value data and typical comparison is presented in Fig. 5(a). It can be seen that recorded shear wave velocity profiles is following almost the same trend as compared to SPT N value.

The lithology is also compared with the shear wave velocity profile developed in these regions using MASW and given as Fig. 5(b). On an average six shear wave velocity profiles are taken near to each of the sites where lithology is known. The average the shear wave velocity profiles obtained in each region is compared with the lithology profiles. It can be seen that the for 100 m thick idealized lithology profiles given by Singh (1996) is compared with the shear wave velocity profiles obtained in the present study. The V_s profile obtained in this study shows the three-distinct region

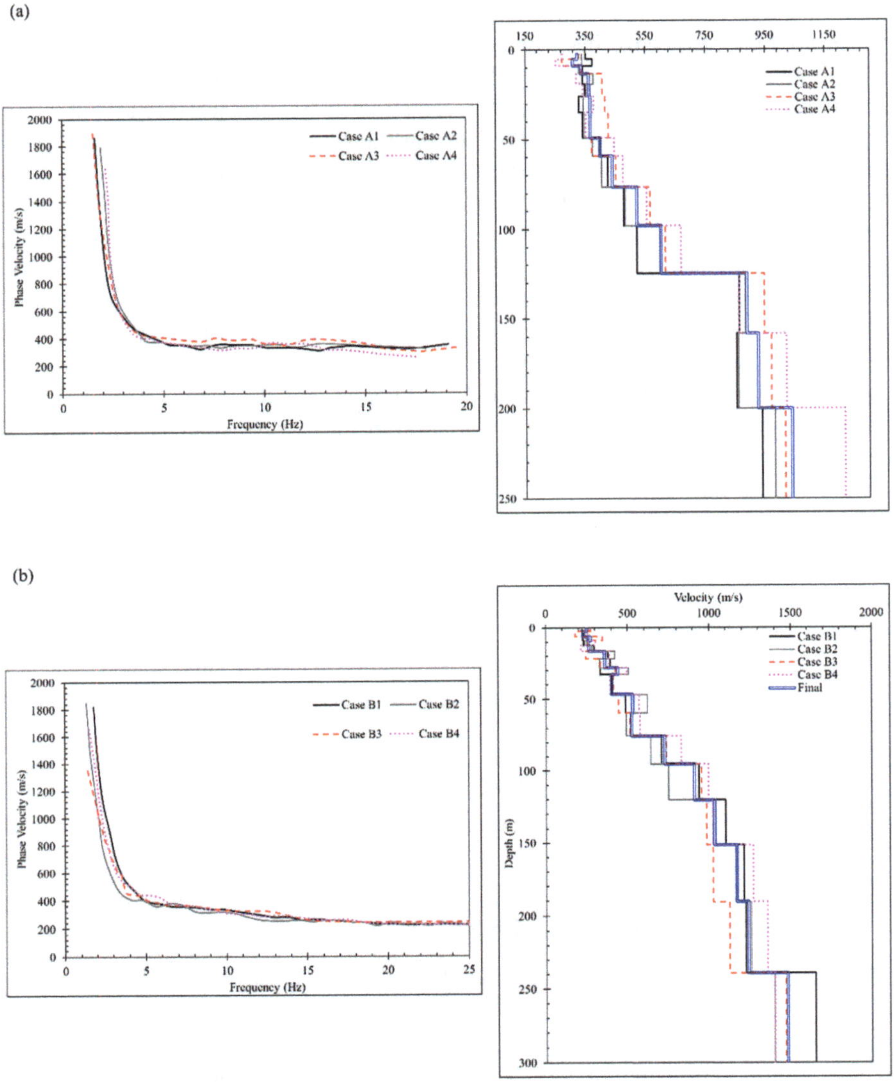

Fig. 4. Typical example of variation of dispersion curve and shear wave velocity profile for variability in (a) sample interval and (b) time of recording data acquisition

(a) a high gradient, low-velocity near-surface region, (2) an intermediate region with shear wave velocities of about 400–450 m/s, and (3) a high-velocity region at depth with velocities increasing to 600–800 m/s.

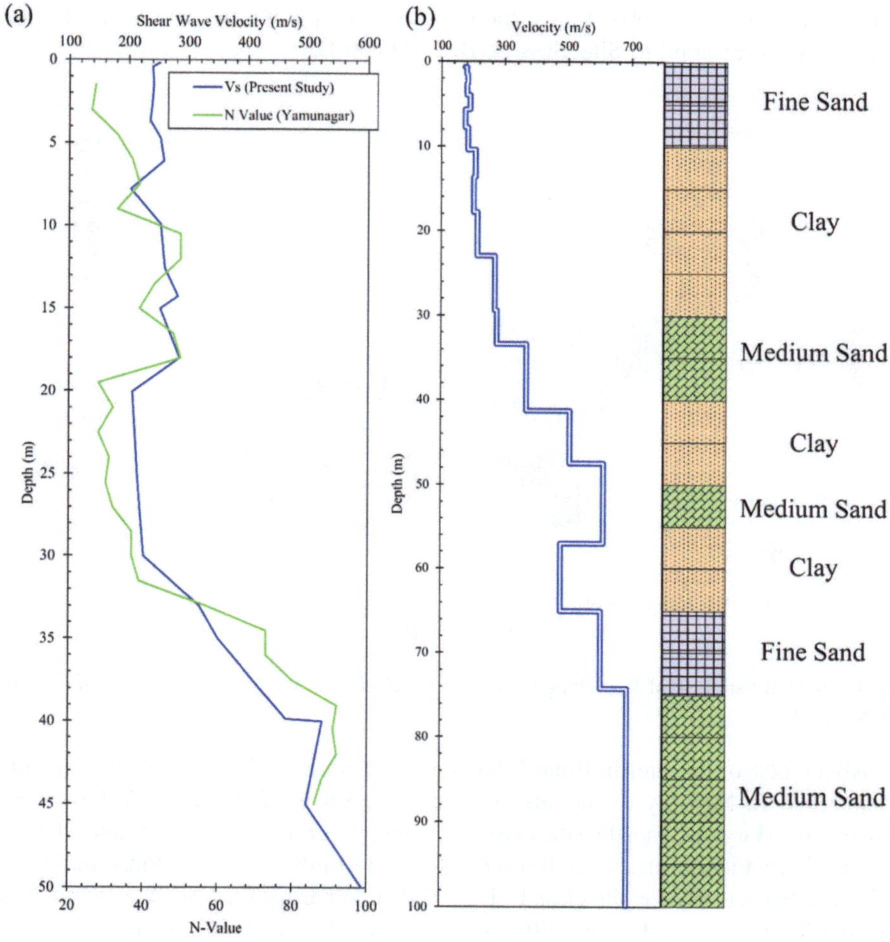

Fig. 5. Typical variation of shear wave velocity with (a) SPT-N value and (b) lithology

5.3 Seismic Site Classification

The shear wave velocity profiles have been developed for 275 sites in Indo Gangetic Basin. The time-averaged shear wave velocity at top 30 m has been calculated using

$$V_{s30} = \frac{\sum_{1}^{N} d_i}{\sum_{1}^{N} \frac{d_i}{V_{si}}} \tag{1}$$

where, d_i is the thickness of layer i and V_{si} is the shear wave velocity of the layer i and V_{s30} is the time average shear wave velocity in the top 30 m. Time average shear wave velocity up to 30 m depth has been calculated for all the entire IGB. Figure 6 shows the variation of V_{s30} along the IGB. The sites have been classified as per National Hazard Reduction Program (NEHRP, BSSC 2003). Profiles with $V_{s30} > 1500$ m/s,

$760 < V_{s30} \leq 1500$ m/s, $360 < V_{s30} \leq 760$ m/s, $180 < V_{s30} \leq 360$ m/s and $V_{s30} < 180$ m/s respectively correspond to Site class A, B, C, D and E.

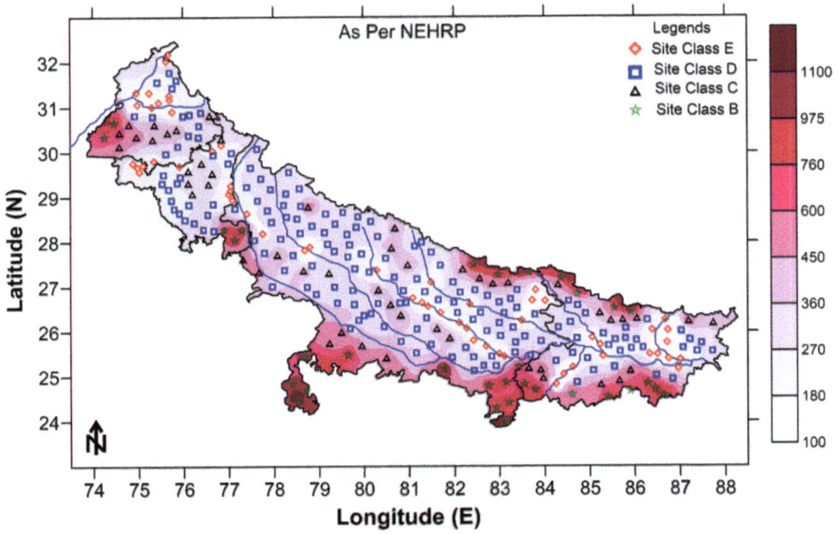

Fig. 6. Spatial variation of V_{s30} along the entire stretch of IGB with seismic site classification as per NEHRP

About 22% of the sites in Bihar is having V_{s30} between 153–180 m/s and classified as site class E. Majority of the sites in Uttar Pradesh is of site class D, however in Punjab and Haryana, mostly site class are either C or D. For Punjab and Haryana region, V_{s30} varies from 160 to 180 m/s, 206 to 340 m/s, 382 to 620 m/s and 795 to 1251 m/s respectively for site class E, D, C and B. For Uttar Pradesh, V_{s30} varies from 157 to 180 m/s, 247 to 358 m/s, 401 to 630 m/s and 822 to 1136 m/s respectively for site class E, D, C and B. For Bihar region V_{s30} varies from 153 to 180 m/s, 214 to 354 m/s, 387 to 615 m/s and 795 to 1152 m/s respectively for site class E, D, C and B.

Further the spatial variation of V_{s30} is compared with the existing geological map of the IGB (See Fig. 1). Most of the sites near to the Kosi, Satluj and Gomti river is of seismic site class E; this may be due to the active sedimentation since Quaternary. Channel alluvium is made up of medium to coarse sand near to Ghaggar river, hence V_{s30} variation is from 180 to 360 m/s. Further near to Sutlej river, low velocity is observed which is due to the presence of silt and clay along its territory. Most of the area of Uttar Pradesh having V_{s30} between 270 to 450 m/s. It is due to the presence of Varanasi Older Alluvium of the Quaternary sediments followed disconformably by Newer alluvium. Most of the sites in the southwest side of Yamuna in UPR is of site class B, which may be due to the presence of Bundelkhand Faridabad ridge and Archean granites and Precambrian Metamorphic (See Fig. 1). Similarly, due to the presence of Munger-Saharsa Ridge and Paleozoic granite in the south of Ganga in Bihar, most of the sites are either site class B or C. Many of the sites in northeast side of Uttar Pradesh and Bihar region is of site class B and C, which may be due to the presence of Siwalik Hills (See Fig. 1).

It can be noticed here that even though V_{s30} is widely used for seismic site characterization, but soil depth extends beyond 100 m. So, the applicability of V_{s30} moreover, its connection with actual amplification need to be examined for IGB region.

6 Conclusions

In addition to earthquake magnitude and its distance from the epicentre, soil characteristic places a major role in the extent of damages at any site due to any seismic event. In this study, an attempt has been made to determine the shear velocity profile up to 400 m depth using combined active and passive multichannel analysis of surface wave survey in 275 selected locations in IGB. Acquisition and processing of the obtained data has been explained to get the V_s profiles at shallower as well as deeper depths. Parametric study has been done to identify the variation in dispersion and V_s profiles by recording time, different sampling interval, different polarity and gain. On an average the difference of 5 to 50 m/s is observed for shallow depth and 150 to 250 m/s at deeper depths in V_s It has seen see that changing any parameter during data recording may result in variation in V_s. Passive recording need to be done for more than 60 s and to combined active and passive data should be collected in case of low background noise. Further, these sites are classified and characterized based on time-averaged V_s in the upper 30 m depth as per NEHRP seismic site classification. For Punjab and Haryana region, V_{s30} varies from 160 to 180 m/s, 206 to 340 m/s, 382 to 620 m/s and 795 to 1251 m/s respectively for site class E, D, C and B. For Uttar Pradesh, V_{s30} varies from 157 to 180 m/s, 247 to 358 m/s, 401 to 630 m/s and 822 to 1136 m/s respectively for site class E, D, C and B. For Bihar region V_{s30} varies from 153 to 180 m/s, 214 to 354 m/s, 387 to 615 m/s and 795 to 1152 m/s respectively for site class E, D, C and B. This is the first time such extensive study was carried out in IGB for V_s determination at deep and shallow depths.

Acknowledgment. The authors thank the Science and Engineering Research Board (SERB) of the Department of Science and Technology (DST), India for funding the project titled "Measurement of shear wave velocity at deep soil sites and site response studies", Ref: SERB/F/162/2015-2016. Authors also thanked to Geological Society of India for providing the Geological and Lithological map required for the study. The authors would like to extend their sincere appreciation to the Deanship of Scientific Research at King Saud University for funding Research group NO. (RG -1435-09).

References

Aki, K.: Space and time spectra of stationary stochastic waves, with special reference to microtremors. Bull. Earthq. Res. Inst. **35**, 415–456 (1957)

Anbazhagan, P., Sitharam, T.G.: Spatial variability of the weathered and engineering bedrock using multichannel analysis of surface wave survey. Pure. appl. Geophys. **166**(3), 409–428 (2008)

Anbazhagan, P., Kumar, A., Sitharam, T.G.: Amplification factor from intensity map and site response analysis for the soil sites during 1999 Chamoli earthquake. In: Proceedings of the 3rd Indian Young Geotechnical Engineers Conference, New Delhi, pp. 311–316 (2010)

Bilham, R., Ambraseys, N.: Apparent Himalayan slip deficit from the summation of seismic moments for Himalayan earthquakes, 1500–2000. Curr. Sci. **88**, 1658–1667 (2005)

Bilham, R., et al.: Himalayan seismic hazard. Sci. **293**, 1442–1444 (2001)

Boominathan, A., Dodagoudar, G.R., Suganthi, A., Maheshwari, R.U.: Seismic hazard assessment of Chennai city considering local site effects. J. Earth Syst. Sci. **117**(S2), 853–863 (2008)

BSSC: NEHRP recommended provision for seismic regulation for new buildings and other structures (FEMA 450). Part 1: provisions, building safety seismic council for the federal Emergency Management Agency, Washington DC (2003)

Desai, S.S., Choudhury, D.: Spatial variation of probabilistic seismic hazard of Mumbai and surrounding region. Nat. Hazards **17**(1), 1873–1898 (2014)

Desai, S.S., Choudhury, D.: Site-specific seismic ground response study for Nuclear power plants and ports in Mumbai. Nat Hazards Rev (2015). https://doi.org/10.1061/(ASCE)NH.15276996.0000177,04015002

Foti, S., Lancellota, R., Socco, L.V., Sambuelli, L.: Application of FK analysis of surface waves for geotechnical characterization. In: Proceedings of the Fourth International Conference on Recent Advances in Geotechnical Earthquake Engineering and Soil Dynamics and Symposium in Honour of Professor W.D. Liam Finn, March 26–31, 2001, San Diego, California, paper no. 1.14, 6 pp (2001)

Humire, F., Sáez, E., Leyton, F., Yañez, G.: Combining active and passive multi-channel analysis of surface waves to improve reliability of V_S30 estimation using standard equipment. Bull. Earthq. Eng. **13**, 1303–1321 (2015)

Jishnu, R.B., et al.: Ground response analysis of Kanpur soil along Indo-Gangetic Plains (2013). https://doi.org/10.1016/j.soildyn.2013.04.001

Kumar, A., Anbazhagan, P., Sitharam T.G.: Site-specific ground response study of deep Indo-Gangetic Basin using representative regional ground motions, Geo-Congress, State of art and practice in Geotechnical Engineering, Oakland, California, paper no. 1065 (2012)

Kumar, A., Bora, O., Harinarayan, N.H.: Obtaining the surface PGA from site response based on globally recorded ground motions and matching with the codal values. Nat. Hazards **81**, 543–572 (2016)

Lyon-Caen, H., Molnar, P.: Gravity anomalies, flexure of the Indian Plate, and the structure, support and evolution of the Himalaya and Ganga Basin. Tecton.S **4**, 513–538 (1985)

Pandey, B., et al.: Site characterization of strong motion recording stations of Delhi using joint inversion of phase velocity dispersion and H/V curve. Bull. Seismol. Soc. Am. (2016). https://doi.org/10.1785/0120150135

Park, C., Miller, R., Xia, J.: Imaging dispersion curves of surface waves on multi-channel record. Society of Exploration Geophysicists Expanded Abstracts 1377–1380 (1998)

Park, C.B., Miller, R.D., Xia, J., Ivanov, J.: Multichannel analysis of surface waves (MASW)-active and passive methods. Lead. Edge **26**, 60–64 (2007)

Park, C., Miller, R.: Roadside passive multichannel analysis of surface waves (MASW). J. Environ. Eng. Geophys. **13**, 1–11 (2008)

Rao, M.B.R.: The subsurface geology of the Indo-Gangetic Plains. Geol. Soc. India. **14**, 217–242 (1973)

Singh, I.B.: Sedimentological history of Quaternary deposits in Gangetic plain. Indian J. Earth Sci. **14**, 272–282 (1987)

Singh, I.B.: Geological evolution of Ganga plain—an overview. J. Palaeontol. Soc. India **41**, 99–137 (1996)

Valdiya, K.S.: Trans-Himadri thrust and domal upwarps immediately south of collision zone and tectonic implications. Curr. Sci. **56**, 200–209 (1987)

Xia, J., Miller, R.D., Park, C.B.: Estimation of near-surface shear-wave velocity by inversion of Rayleigh waves. Geophys. **64**, 691–700 (1999)

Turbulent Flow Characteristics in Interior and Wake Region of Emergent and Sparse Vegetation Patch

Soumen Maji[1,2(✉)], Prashanth Reddy Hanmaiahgari[2], Ram Balachandar[3], and Vesselina Roussinova[4]

[1] Department of Civil Engineering, CIT Kokrajhar, Kokrajhar, Assam, India
s.maji@cit.ac.in
[2] Department of Civil Engineering, IIT Kharagpur, Kharagpur, West Bengal, India
[3] Department of Civil and Environmental Engineering, University of Windsor, Windsor, ON N9B 3P4, Canada
[4] Department of Mechanical, Automotive and Materials Engineering, University of Windsor, Windsor, ON N9B 3P4, Canada

Abstract. Vegetation in the flow field affects flow hydrodynamics. For example, emergent vegetation patch in the flow field produces changes in river morphology, changes in flow velocity and turbulence. Therefore, investigation of flow characteristics in and around the emergent vegetation patch is a topic of importance. The main aim of this study is to investigate the self-similarity and scaling of profiles of streamwise and lateral velocities, turbulence intensities, Reynolds shear stress and kinetic energy in the interior and the wake region of the vegetation patch. A Nortek Vectrinoplus Acoustic Doppler Velocimeter (ADV) was used to measure the point velocity data throughout the depth along the centerline of the vegetation patch in the streamwise direction. Seventy numbers of acrylic cylinders were used to make the emergent vegetation array in the middle of a laboratory flume. In this study, flows with two different Reynolds numbers and two different aspect ratios were investigated. The results demonstrate self-similarity of streamwise velocity profiles normalized by depth averaged velocity in the outer region in the interior and wake region of the emergent and sparse vegetation patch irrespective of Reynolds number and aspect ratio of flow. Similarly, wall wake similarity was observed in the Reynolds shear stress distribution.

1 Introduction

Variation of turbulent flow characteristics along the streamwise centerline of the vegetation patch in the interior and wake region is investigated for two Reynolds numbers and two aspect ratios on a rough bed. Significance of this study is to investigate the similarity of turbulence characteristics in the interior of the vegetation at various flow conditions. Further, scaling of turbulence characteristics throughout the depth is investigated. The non-dimensional similarity profiles can be used in developing analytical solutions to similar vegetated flows.

J. Wasowski and T. Dijkstra (Eds.): GeoMEast 2018, SUCI, pp. 73–82, 2019.
https://doi.org/10.1007/978-3-030-02032-3_7

Fox and Stewart (2014) carried out mixed scaling and similarity analysis for turbulent open-channel flow over gravel and cobble beds using the asymptotic invariance principle (AIP). Cheng et al. (2012) investigated scaling of velocity profile and velocity defect in sparse submerged rigid vegetation canopy, in which vertical height is scaled by height of the vegetation, and proposed a power law for the velocity profile with exponent limited between 2.5 and 9. Tang and Hsieh (2015) developed analytical solutions for the velocity profiles, shear stress profiles and kinetic energy dissipation rates in submerged vegetation as well as emergent vegetation low Reynolds number flows using non-dimensional governing equations. Maji et al. (2016, 2017) studied channel flow with emergent vegetation patch, mainly focusing on streamwise velocity variation along the transverse direction. Meftah and Mossa (2016a, b) proposed a log law for the depth averaged streamwise velocity in the shear layer adjacent to the patch as a function of transverse distance from the edge of the patch. However, Meftah and Mossa (2013, 2016a, b) study is limited to the transversal flow velocity profile influenced by momentum transfer in the form of shear stress at the interface between the array and the unobstructed flow. Kubrak et al. (2008) developed a one dimensional model to study the velocity profile for submerged and emergent vegetation using the mixing length. Huai et al. (2009a, b) proposed a numerical model for submerged and emergent rigid vegetation using von Kármán similarity.

In this study, special attention is given to the similarity of velocity and turbulence quantities in the interior of a sparse and emergent vegetation patch which is lacking in the literature. The content of this study is arranged as follows. Experimental setup and methodology of experimental data collection at various streamwise locations in the upstream, interior and downstream of the sparse and emergent vegetation patch to investigate hydrodynamics is explained in experimental methodology section. In the results and discussion, similarity of velocity profiles, turbulence intensities, Reynolds shear stress have been investigated along the streamwise centerline of vegetation patch and the wake region. Finally, results and discussion are summarized in conclusions section.

2 Experimental Methodology

All experiments were performed in a wide rectangular channel (12 m long, 0.91 m wide and 0.61 m deep) existing in the Hydraulic and Water Resources Engineering Laboratory of Indian Institute of Technology Kharagpur, India. The flume contained two glass sidewalls (8 m long and 15 mm thickness) for the clear view of the flow. The test section of the flow was considered as 3 m long which begins from 7 m downstream of the inlet. The water depths (h) in the different experiments were maintained as 150 mm and 200 mm by a tailgate which was fixed at the downstream end of the flume. The maintained average flow velocity was 0.29 m/s and 0.31 m/s respectively for different experiments. The cross-sectional lines through those streamwise locations were located in the upstream (x = 0 m, 0.50 m, 0.90 m, 1.00 m), interior (x = 1.30 m, 1.50 m, 1.78 m, 1.90 m) and downstream (x = 2.20 m, 2.50 m, 2.85 m and 3.00 m) of the vegetation patch. Figure 1a shows the photograph of vegetation patch placed in the middle of the flume. Array of acrylic circular rods was used to simulate rigid and emergent vegetation patch. To prevent the vibrations, the acrylic rods top ends were properly fixed. Table 1 gives the complete details of the vegetation patch.

(a)

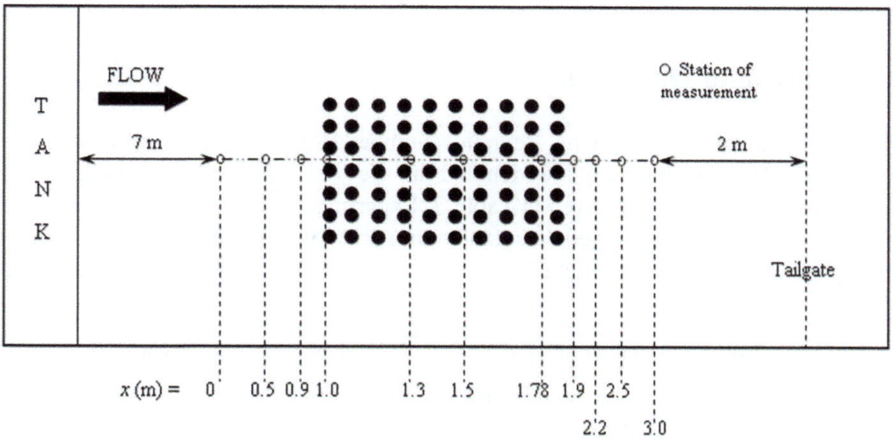

(b)

Fig. 1. (a) Photograph of vegetation patch and Acoustic Doppler Velocimetry (flow direction is from bottom to top). (b) Top view schematic of streamwise measurement locations

In this study, turbulence along the centerline of the vegetation patch which is 0.474 m from the RHS flume wall (RHS is with reference to the looking in the flow direction) has been investigated (refer Fig. 1b). Refer to Fig. 1b for (not to scale) top view of the streamwise locations where experimental data were measured. Velocities

Table 1. Details of vegetation patch

Vegetation length (x)	0.816 m
Vegetation width (y)	0.246 mm
Rods per unit area	350 rods/m^2
Length of vegetation cylinder	300 mm
Diameter of vegetation cylinder	6.4 mm
Solid volume fraction	0.00894

were measured with the Acoustic Doppler Velocimetry. Flow depths were determined by a graduated millimeter scale with an accuracy ±0.1 mm.

Velocity was measured at fifteen vertical locations which were z = 3, 5, 7, 9, 15, 20, 25, 30, 40, 50, 60, 70, 80, 90 and 100 mm for water depth 150 mm (in addition 120 mm and 150 mm for water depth 200 mm) from the channel bed. Data was recorded at each vertical location for 300 s with a sampling frequency of 100 Hz and sampling volume of 2.5×10^3 mm^3. From the velocity data recorded, random spikes were removed by the spike filtering technique as suggested by the previous researchers (Goring and Nikora 2002; Wahl 2003; Mori et al. 2007). The removed data was replaced by the values obtained from cubic interpolation. To ensure the quality of the data obtained after spike filtering, the signal-to-noise ratio test as proposed by Chanson et al. (2008) was implemented. The signal-to-noise ratio was maintained as 15, and the data with correlation values greater than 75% were considered.

3 Results and Discussion

Three experiments with varying Reynolds number (Re) and aspect ratio (Ar) were conducted. The experiments are denoted as Re1Ar1, Re1Ar2 and Re2Ar2. All experimental conditions are given in Table 2. Experimental data, including interior and wake region were plotted together to study similarity of important flow variables and other parameters of the flow field for the varying Reynolds number and aspect ratio.

Table 2. Details of flow conditions in the laboratory experiments

Exp.	h (mm)	$\bar{u}$ (m/s)	u_{max} (m/s)	u_* (m/s)	Mean flow R_e	F_r	A_r
Re1Ar1	155	0.29	0.30	0.238	43500	0.235	5.87
Re1Ar2	200	0.22	0.25	0.197	43500	0.157	4.55
Re2Ar2	200	0.31	0.34	0.254	62000	0.221	4.55

3.1 Similarity of Velocity Profiles

Distributions of streamwise (u) and lateral velocities (v) are scaled with depth averaged velocity ($\bar{u}$) and velocity vector $\left(\sqrt{v^2 + w^2}\right)$ in the lateral direction respectively to study the self-similarity of velocities in the interior and the wake region of the vegetation patch. Henceforth, the scaled streamwise velocity profiles are plotted in Fig. 2a and b for interior and wake regions respectively. The streamwise velocity profile is uniform and rectangular in shape in the interior region (Fig. 2a), whereas velocity profile follows a power law in the wake region (Fig. 2b). In the interior of the vegetation patch similarity of the streamwise velocity profile is observed in the outer layer but not in the inner layer (Fig. 2a). For a given Reynolds number, with an increase in aspect ratio, velocity profile is attaining a uniform velocity throughout the depth within a short distance from the leading edge. Similarly, for a given aspect ratio, with an increase in Reynolds number, velocity profile becomes uniform throughout the depth within a short distance from the leading edge. Thus, there is a disagreement of velocity profiles in the inner layer at different streamwise locations in the interior of the vegetation patch.

In the wake region of vegetation patch, similarity of the streamwise velocity profile is observed in the outer layer but not in the inner layer because velocity profile is trying to attain the fully developed velocity profile from the rectangular velocity profile as the distance increases in the downstream direction. For a given Reynolds number, with an increase in aspect ratio, the distance required for rectangular velocity profile to attain fully developed velocity profile is becoming shorter. Similarly, for a given aspect ratio, with an increase in Reynolds number, the downstream distance required for rectangular velocity profile to attain fully developed is decreasing. Thus, there is a disagreement of velocity profiles in the inner layer at different streamwise locations in the wake of the vegetation patch.

It is observed from Fig. 2c and d that lateral velocity vector in the wake region follows the same trend as in the interior of the vegetation. The scatter in the data is also similar in the interior and wake regions. The scaled lateral velocity in the interior of the vegetation is increasing in the streamwise direction, whereas the scaled lateral velocity in the wake region is decreasing with increase in the streamwise distance. In addition, the scaled lateral velocity is increasing with increase in the aspect ratio. The probable reason behind this characteristic is that the intensity of lateral flow increases along the centerline of the vegetation patch as the flow increasingly deviates from the central region towards the sidewalls, owing to the resistance offered by the vegetation patch, however, the strength of the lateral flow is decreasing in the wake region along the downstream direction because of diminishing wake effect along the streamwise direction.

3.2 Turbulence Intensities

Upstream of the vegetation patch, the flow along the centerline is irrotational and slows down due to the blocking effect of the entire patch and the drag force from the cylinders. The flow bleeds through the vegetation with low velocity, and high-velocity flow passes through the region without vegetation *i.e.*, the region between sidewalls and vegetation patch. Distributions of turbulence intensities (u', v' and w') are scaled with RMS value of kinetic energy $\left(\sqrt{k}\right)$ in the interior and wake region for all the experimental runs have been shown in Fig. 3(a, b, c, d, e and f). Observing the Fig. 3, it has been concluded that distributions of turbulence intensities in the interior and wake

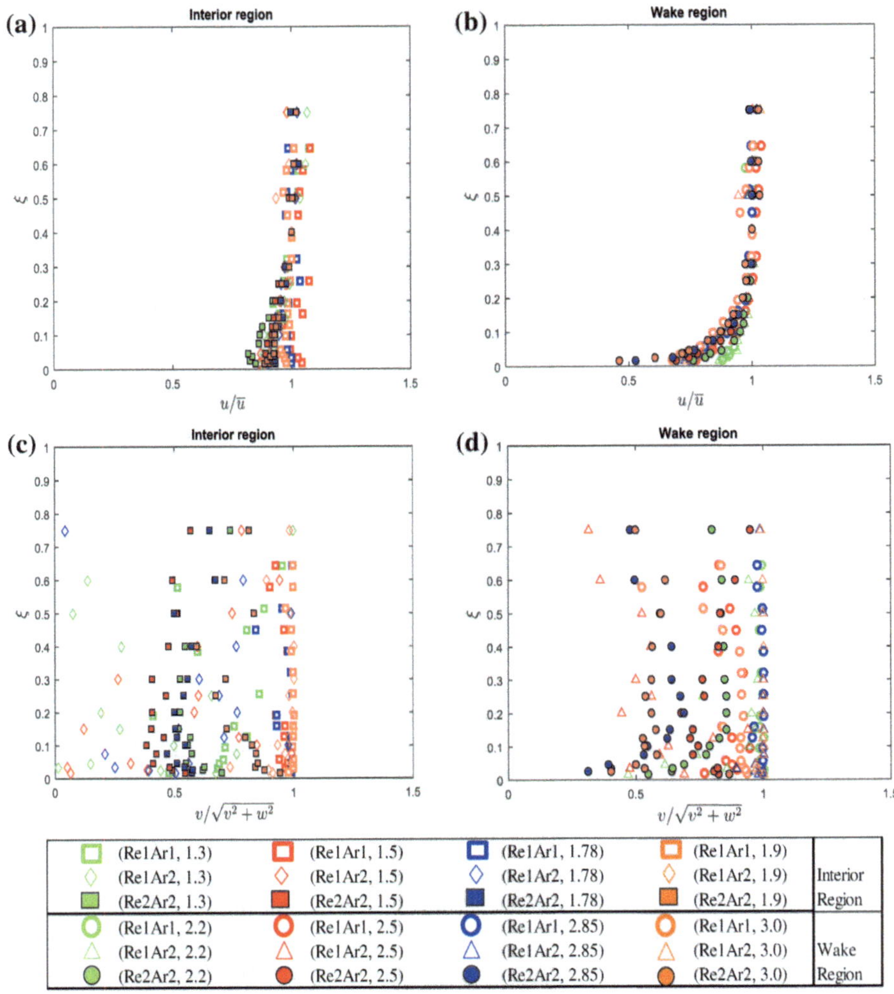

Fig. 2. Streamwise velocity distribution profiles throughout depth (a) interior and (b) wake region, and lateral velocity vectors throughout depth (c) interior and (d) wake region

region along the centerline of the patch in the streamwise direction follow the same geometry. Change in Reynolds number and aspect ratio found to be affecting the turbulence intensity profiles. In Fig. 3a, streamwise turbulence intensity is decreasing in the streamwise direction. Moreover, the streamwise turbulence intensity in the interior of the vegetation found to be increasing with increasing aspect ratio and Reynolds number. Streamwise turbulence intensity variation in the wake region is shown in Fig. 3b. In the wake region streamwise turbulence intensity is increasing in the downstream direction. Similar to the interior region, the streamwise turbulence intensity in the wake region is found to be increasing with increasing aspect ratio and Reynolds number. Streamwise turbulence intensity is the maximum near the bed and decreasing towards the free surface in the interior and the wake region. In contrast, lateral and vertical turbulence intensities are increasing with increasing vertical coordinate. The lateral turbulence intensity in the interior of the vegetation is increasing in the downstream direction which is contrary to the behavior of streamwise turbulence intensity (Fig. 3c). The lateral turbulence intensity in the interior of the vegetation patch is increasing with increasing aspect ratio. The lateral turbulence intensity profile in the wake region is similar to that in the interior of the vegetation patch. The lateral turbulence intensity in the wake region is decreasing with increase in the streamwise distance. Similar to the behavior in the interior of the patch, the lateral turbulence intensity in the wake region is increasing with increasing aspect ratio. Vertical turbulence intensity profiles in the interior and wake regions are shown in Fig. 3e and f. Vertical turbulence intensity inside the vegetation patch and in the wake region is independent of the streamwise location, however, w' is increasing with increase in Reynolds number.

3.3 Reynolds Shear Stress

Reynolds shear stress $-\overline{u'w'}$ plays a dominant role in and around the vegetation patches compared to other two Reynolds shear stresses *i.e.* $-\overline{u'v'}$ and $-\overline{v'w'}$. Therefore, vertical distributions of Reynolds shear stress $-\overline{u'w'}$ along the midsection of the channel for interior of the vegetation and wake region have been shown in Fig. 4(a) and (b). In these plots, maximum Reynolds shear stress $\left(\left(-\overline{u'w'}\right)_{max}\right)$ is used for scaling the Reynolds shear stress. The self-similarity of the RSS profile is poor in the interior of the vegetation patch as shown in Fig. 4a. The magnitude of $-\overline{u'w'}$ in the interior vegetation (Fig. 4a) is more than the wake region (Refer Fig. 4b) due to intense turbulent mixing and high interaction between the vortices in the interior of the vegetation patch. In the interior of vegetation patch, Reynolds shear stress $-\overline{u'w'}$ decreases as the distance increases in the downstream and this trend continues up to the trailing edge of the vegetation patch. In the interior of the vegetation, the Reynolds shear stress $-\overline{u'w'}$ is found to be increasing with increasing aspect ratio and Reynolds number. In the interior of the vegetation patch, peak Reynolds shear stress occurring away from the bed which shows higher thickness of roughness sublayer in the interior.

In the wake region, the Reynolds shear stress profile follows an exponential curve which decreases as the vertical coordinate increases as shown in Fig. 4b. The self-similarity of the Reynolds shear stress in the wake region is better than that in the interior of the vegetation. In the wake region of vegetation patch, Reynolds shear stress $-\overline{u'w'}$ increases as the distance increases in the downstream as the flow is recovering from the wake created by the porous vegetation patch. In the wake region, the peak of RSS occurs close to the bed which indicates thin roughness sublayer.

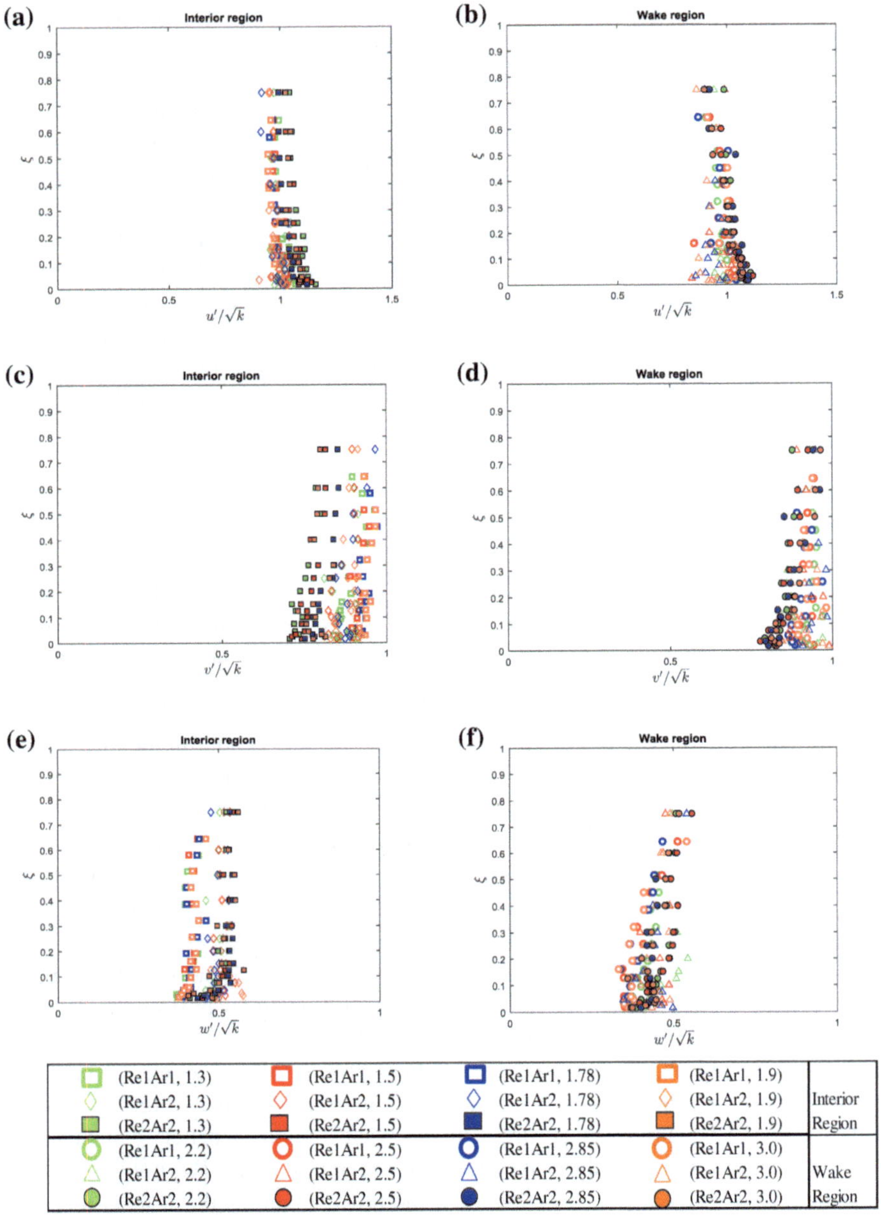

Fig. 3. Streamwise turbulence intensity profile along depth in (a) interior vegetation patch and (b) wake region; lateral turbulence intensity profile in (c) interior and (d) wake region; and vertical turbulence intensity in (e) interior and (f) wake region

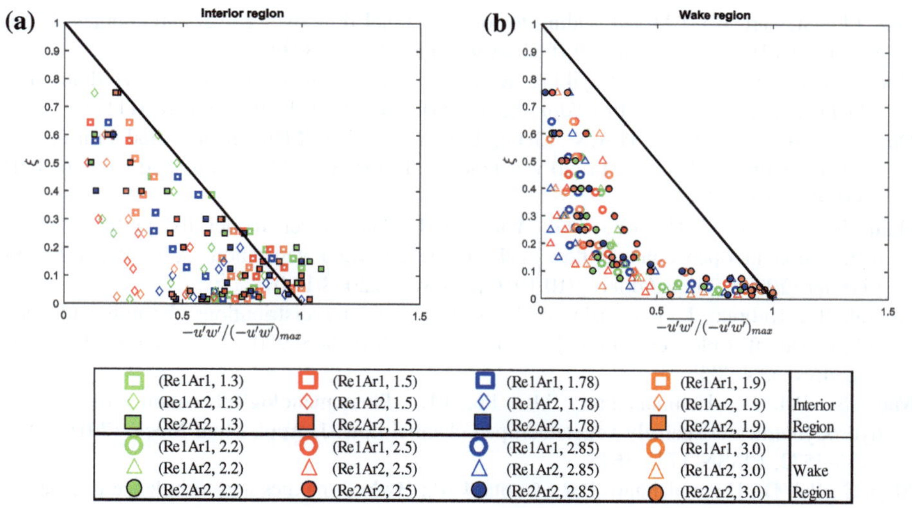

Fig. 4. Vertical distribution of Reynolds shear stress in (a) interior and (b) wake region (Thick line shows theoretical linear distribution in uniform open channel flow)

4 Conclusions

Similarity of streamwise velocity profile scaled by depth average velocity is observed in the outer layer in the interior and wake of the vegetation patch irrespective of aspect ratio and Reynolds number. The similarity of the lateral velocity profile scaled by the lateral velocity vector is not very good. The scaled lateral velocity in the interior of the vegetation is increasing in the streamwise direction, whereas the scaled lateral velocity in the wake region is decreasing with increasing streamwise distance.

Self-similarity is observed in the turbulence intensities scaled by kinetic energy. The streamwise, lateral and vertical turbulence intensities in the interior and wake region are found to be increasing with increase in aspect ratio and Reynolds number.

Reynolds shear stress is normalized by shear velocity. The self-similarity of the Reynolds shear stress in the wake region is better than that in the interior of the vegetation. In the wake region, the Reynolds shear stress profile follows an exponential curve with maximum near the bed. Reynolds shear stress is found to be increasing with increase in the Reynolds number.

References

Chanson, H., Trevethan, M., Aoki, S.I.: Acoustic Doppler Velocimetry (ADV) in small estuary: field experience and signal post-processing. Flow Meas. Instrum. (2008). https://doi.org/10.1016/j.flowmeasinst.2008.03.003

Cheng, N.-S., Nguyen, H.T., Tan, S.K., Shao, S.: Scaling of velocity profiles for depth-limited open channel flows over simulated rigid vegetation. J. Hydraul. Eng. (2012). https://doi.org/10.1061/(ASCE)HY.1943-7900.0000562

Fox, J.F., Stewart, R.L.: Mixed scaling for open-channel flow over gravel and cobbles. J. Eng. Mech. (2014). https://doi.org/10.1061/(ASCE)EM.1943-7889.0000793

Goring, D.G., Nikora, V.I.: Despiking acoustic Doppler velocimeter data. J. Hydraul. Eng. **128**(1), 117–126 (2002). https://doi.org/10.1061/(asce)0733-9429(2002)128:1(117)

Huai, W.X., Chen, Z.B., Jie, H.A.N., Zhang, L.X., Zeng, Y.H.: Mathematical model for the flow with submerged and emerged rigid vegetation. J. Hydrodyn. (2009a). https://doi.org/10.1016/s1001-6058(08)60205-x

Huai, W.X., Zeng, Y.H., Xu, Z.G., Yang, Z.H.: Three-layer model for vertical velocity distribution in open channel flow with submerged rigid vegetation. Adv. Water Resour. Elsevier (2009b). https://doi.org/10.1016/j.advwatres.2008.11.014

Kubrak, E., Kubrak, J., Rowiński, P.M.: Vertical velocity distributions through and above submerged, flexible vegetation. Hydrol. Sci. J. **53**(4), 905–920 (2008). https://doi.org/10.1623/hysj.53.4.905

Maji, S., Pal, D., Hanmaiahgari, P.R., Pu, J.H.: Phenomenological features of turbulent hydrodynamics in sparsely vegetated open channel flow. J. Appl. Fluid Mech. (2016). ISSN 1735-3572, EISSN 1735-3645

Maji, S., Pal, D., Hanmaiahgari, P.R., Gupta, U.P.: Hydrodynamics and turbulence in emergent and sparsely vegetated open channel flow. Environ. Fluid Mech. **17**, 853–877 (2017). https://doi.org/10.1007/s10652-017-9531-2

Meftah, M.B., Mossa, M.: Prediction of channel flow characteristics through square arrays of emergent cylinders. Phys. Fluids (2013). https://doi.org/10.1063/1.4802047

Meftah, M.B., Mossa, M.: A modified log-law of flow velocity distribution in partly obstructed open channels. Environ. Fluid Mech. (2016a). https://doi.org/10.1007/s10652-015-9439-7

Meftah, M.B., Mossa, M.: Partially obstructed channel: contraction ratio effect on the flow hydrodynamic structure and prediction of the transversal mean velocity profile. J. Hydrol. **542**, 87–100. (2016b). https://doi.org/10.1016/j.jhydrol.2016.08.057

Mori, N., Suzuki, T., Kakuno, S.: Noise of acoustic Doppler velocimeter data in bubbly flows. J. Eng. Mech. (2007). https://doi.org/10.1061/(asce)0733-9399(2007)133:1(122)

Tang, C.Y., Hsieh, P.C.: Dynamic analysis of vegetated water flows. J. Hydrol. Eng. (2015). https://doi.org/10.1061/(ASCE)HE.1943-5584.0001294

Wahl, T.L.: Discussion of "Despiking acoustic doppler velocimeter data" by Derek G. Goring and Vladimir I. Nikora. J. Hydraul. Eng. (2003). https://doi.org/10.1061/(asce)0733-9429(2003)129:6(484)

Volcanic Disaster and the Decline of Mataram Kingdom in the Central Java, Indonesia

Sari Bahagiarti Kusumayudha(✉), Helmy Murwanto, Sutarto, and Siti Umiyatun Choiriyah

Universitas Pembangunan Nasional Veteran Yogyakarta, Yogyakarta, Indonesia
saribk@upnyk.ac.id

Abstract. In the area of Central Java and Yogyakarta there are several volcanoes namely Sindoro volcano, Sumbing volcano, Slamet volcano, Ungaran volcano, Merbabu volcano, and Merapi volcano. Among the volcanoes, Merapi volcano is the most active, generating lava flows, pyroclastic flows, glowing clouds, and lahar. On the other hand, during the 7th to 10th century in this region had been existed an ancient monarchy, known as the Mataram Kingdom. Its glory at that time was marked by many ancient heritages especially in the form of temples. There are many monumental, beautiful, and majestic archaeological relics, located on the plains, slopes, even near the summit of Merapi volcano, Sumbing volcano, Sindoro volcano, Dieng mountains, and Ungaran volcano.

In the 11th century the history of Mataram Kingdom was not recorded anymore, suspected that the kingdom declined, and the cultural center of excellence transferred to East Java. Up to now, the cause of the collapse of Mataram Kingdom is still a mystery. Some historians suspect as a result of a great war, and some others thought as the impact of volcanic disaster. Nevertheless, the fall of Mataram Kingdom still being an enigma.

At the southern slopes of Merapi volcano, there are some ancient buildings that buried by volcanic deposits. For example Kadisoka temple part of the building is still dotted sandstone, tuff, and pyroclastic sandstone, cobbles. Similarly, Sambisari, Kedulan, and Pustakasala temples, when discovered, they were sinked by sand, tuff, and fluvio-volcanic sediments. On the slopes of Sindoro volcano, there are also ancient sites, namely Liyangan was covered by pyroclastic flow, tuff, and lahar deposits. Thus it is possible that volcanic disasters have contributed to the decline of Mataram Kingdom.

Keywords: Volcanic disaster · Pyroclastic flow · Lahar · Ancient temples Mataram Kingdom

1 Introduction

In the Central Java and Yogyakarta Special Territory of Indonesia, during the 7th to 10th century, there was an independent state, namely the Mataram Kingdom. The past kingdom glory is marked by the existing of many ancient relics and buildings, two of which are internationally well-known, as Borobudur temple and Prambanan temple. Some beautiful and magnificent monumental archaeological remains have been found

J. Wasowski and T. Dijkstra (Eds.): GeoMEast 2018, SUCI, pp. 83–93, 2019.
https://doi.org/10.1007/978-3-030-02032-3_8

in the study area. The beauty and splendor of the ancient heritage illustrates the taste of architectural artworks of noble, in the form of Buddhist and Shiwa Hinduist temples. The temples are generally scattered on the plains, the feet slopes, the central slopes, the upper slopes, and some even on the summit of the mountains. At the time of its discovery, several temples were heavily damaged or buried by soil and rocks as deep as 2 to 6 m.

There were no archaeological relic discoveries after the 10th century, indicating that in the Central Java did not exist any governmental or state. This was a marker of the sudden disappearance of the ancient Mataram Kingdom from historical records. It was subsequently replaced by the emergence of various kingdoms in the eastern Java, starting from the kingdoms of Singosari, Kediri, Jenggala, to Majapahit, from the 12th century to the 14th century. The kingdom history seemed to be allegedly migrated to the East Java.

According to historians, the movement of the central government from the Central to the East Java was driven by political, cultural and religious aspects. In the matter of facts, some ancient relics were found in a collapsed condition, or buried by sediments of natural works. Therefore, in the framework of providing historical corrections and information, geological-based studies was done to unravel the enigma of the Mataram decline and its relationship with natural disaster in the Central Java and Yogyakarta Special Territory.

Objectives of this study are to mapping and analyzing the geological disaster tracks and records such as volcanic eruption, lava, and ash deposits around ancient archaeological sites that suspected of contributing to the damage the temples. The location of the study area is shown in Fig. 1.

2 Literature Reviews

2.1 History of the Mataram Kingdom

The Mataram Kingdom in the Central Java, established from the 7th to the 10th century, was ruled by kings from the Sanjaya dynasty and the Syailendra dynasty. It is written in the Mantyasih inscription (907 AD), that the Ancient Mataram Kingdom authority was covering Central Java and Yogyakarta territories. But in the late 10th century the royal capital moved to the East Java, and the cultural center of Central Java seemed to shift to East Java, marked by the emergence of Medang, Kahuripan, Kediri, Jenggala, Singosari, and Majapahit kingdoms. The history then returned to the Central Java after Majapahit Kingdom collapsed, entering the Islamic era, starting from Demak Kingdom, Pajang Kingdom, and Islamic Mataram Kingdom founded by Panembahan Senopati (Kusumayudha 2006a, b).

Poesponegoro and Notosusanto (2008) stated that at the time of the Mataram Kingdom under the king Rakai Sumba Dyah Wawa which ruled in 919–925 AD (inscription Air Kali, inscription Kinawe, 928 AD, inscription Kambang Sri Jebung, and inscription Biota), the kingdom undergone a great natural event. The continuity of the government ended suddenly. This is allegedly due to the eruption of Merapi volcano resulting in earthquakes and lahar flooding. The event is expected to result in the

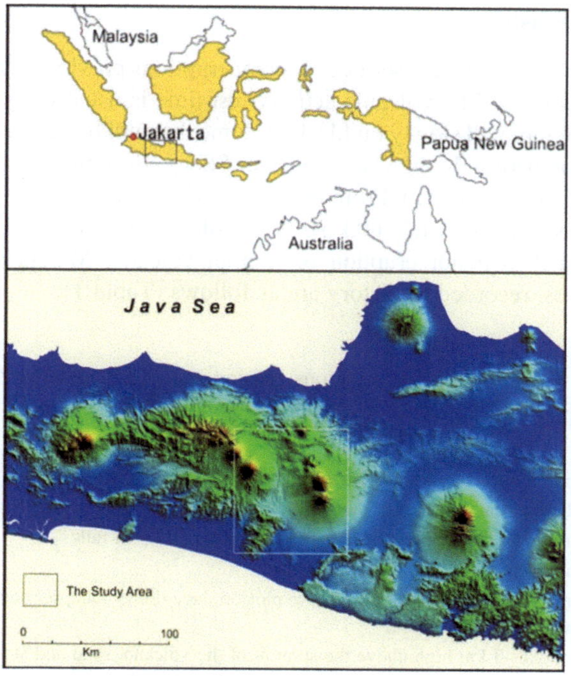

Fig. 1. Map showing the location of the study area

royal center being moved to East Java. In another inscription it is written that the transfer of government occurred in 930 AD, during the reign of the last king of the Mataram Kingdom, Mpu Sindok (928–947 AD). Mpu Sindok subsequently established Medang Kingdom where the center of his government was initially located in the upstream of River Brantas (Poesponegoro and Notosusanto 2008).

According to Van Bemmelen (1949), Neuman van Padang (1951) in Kusumayudha (2006a, b) and MacDonald (1972), Merapi volcano erupted catastrophically in 1006. This eruption resulted in a total paralysis of the economy, farmland severely damaged, and volcanic deposits 6 meters thick covering the ground. It was illustrated that the people and the royal family were destroyed by the lahar floods and buried by volcanic ash.

Nevertheless, this hypothesis was opposed by some volcanologists. Based on inscription Pucangan, during the reign of Dharmawangsa (1016–1017 AD), the kingdom suffered a major catastrophe causing the capital to be destroyed, many deaths, and economical paralyzed. In the event, the cousin of the king, named Airlangga was able to escape to the Southern Mountains. Airlangga then develop the Singosari Kingdom. This happening is known as the *Maha Pralaya* (Zen 2006). With regard to the greatest of praws, related to the decline of the Mataram Kingdom, there are still two different opinions, whether due enemy attacks from the Wura-wari Kingdom (Sri Wijaya) or by the eruption of Merapi volcano (Kusumayudha 2006a, b).

2.2 Volcanic Disaster

Kusumadinata (1979) recorded a series of Merapi eruptions proving that the volcano has been very often erupted. The volcanic activity rest time is usually 5 years. During the eruption break, Merapi will slowly build a lava dome that will be destroyed and ruptured in the next eruption period. In the valley of River Sileng, Borobudur area, there is found such volcanic ash as thick as 2 m. From the results of radio-dating testing, the sediment was estimated to be approximately 600–700 years old. Some evidences show that at that time there was such a major eruption of Merapi volcano. Merapi eruptions which claimed many lives, recorded in history are as follows (Table 1):

Table 1. The records of Merapi eruptions causing lost (Kusumayudha 2013a, 2013b)

Year	Eruption characteristics	Number of life lost
1672	Producing pyroclastic flows, glowing clouds, and lahar	300
1930–1931	Normal eruptions, lava flows, pyroclastic flows, glowing clouds, and lahar	1369
1954	Producing pyroclastic flows, glowing clouds, ash and lapilli falls	64
1961	Producing lava flows and glowing clouds	6
1969	Explosive eruption, glowing cloud of explosion, lava dome collapse, bomb and block falls, and ash rain	3
1972–1973	Black smokes 3 km high above the summit of the volcano, sand and stones falls at the Babadan observatory, lava dome collapse producing pyroclastic flow, glowing clouds to River Batang reaching 3 km distant	
1994	Eruption resulting pyroclastic flows and glowing clouds to the south from the summit to River Boyong, reaching 6 km distance	67
2006	Merapi type eruption, producing pyroclastic flows of 4 km distance to the valley of River Gendol, glowing clouds, and lahar	2
2010	Pelean type eruption, producing lava flows, pyroclastic flows of 17 km distance to the valley of River Gendol, glowing clouds, lapilli and ash falls, lahar	250

The prehistoric major eruptions of Merapi were never known for certain by anyone. Scientists can only try to trace and reconstruct these occurrences with geological, volcanological, historical, archaeological, cultural, even metaphysical approaches. For human life, actually the more important thing is not about when Merapi erupted exactly, but rather how the impact of the eruption to human life (Kusumayudha 2006a, b).

Borobudur temple has a different historical background. Murwanto et al. (2004) found evidences related to the existence of a lake around the temple at that time. According to Van Bemmelen (1949), the ancient Borobudur Lake occurred as a result of a major eruption of Merapi volcano in 1006 AD. This devastating eruption caused apart of the peak of Merapi to collapse to the southwest direction, stemming the flow of River Progo, forming a large lake in the area of South Kedu. The Borobudur great lake as a hollow between the mountains (intermontains basin) was gradually deterred by tectonic and sedimentation processes. In the VIII century. Borobudur temple was built on a small hill surrounded by a lake during the reign of King Smaratungga of the Syailendra dynasty.

The activity of Merapi volcano has a major contribution to the environmental change of the lake and the mainland. The eruption materials gradually hoarded the Borobudur Lake, causing it to become shallow and finally dry at the end of the XIII century. Merapi eruption materials not only dried the lake, but also damaged and buried Borobudur temple as those of other temples in the Central Java and Yogyakarta.

In addition to the primary hazards of pyroclastic flows, glowing clouds and volcanic ash, the secondary threat of volcano that is no less dangerous is lahar. Lahar is a dense stream that occurs when the pyroclastic deposits in the upper slopes of the volcano mixes with rain water becoming saturated, then influenced by gravity to flowing downstream through the river valleys as a heavy currents of mud and rocks (Kusumayudha 2013a, b). Lahar behaves enormous erosion, high destructiveness, capable of transporting very large materials, including boulders, trees, and buildings. Lahar of Merapi in 1930–1931 along with the primary eruption, has claimed of 1369 deaths (Kusumadinata 1979)

3 Methods of Study

The research carried out with descriptive, comparative, and surveys methods that was complemented by geological phenomena mapping. The data used consist of secondary data and primary data. Secondary data includes various information from the existing studies. Primary data obtained through survey and field mapping. Geological studies to be carried out include geomorphology, petrology, stratigraphy, sedimentology, geological structure, and disaster analyses.

The stages to be carried out during the research were divided into three steps. Firstly, to collect secondary data consist of map showing the distribution of ancient relics, regional geological map, tectonic map, disaster map, aerial photography/satellite imagery, and historical research data ever undertaken by other researchers. Secondly, to analyse the geological map, tectonic map, and disaster map as well as remote sensing interpretation using aerial photography/satellite imagery against the spread of archaeological sites, and the presence of geological structures around the sites. Thirdly, to mapping geological details around the archaeological sites, including the dissemination of sediment produced by volcanic activity, either the primary activity of pyroclastic or the secondary activity of lahar.

4 Results and Discussion

4.1 Archeological Sites Discoveries

Some monumental relics sites of the ancient Mataram Kingdom were formerly found in a collapse conditions or buried by soil. Remnants of the Mataram Kingdom glory can be traced from the discovery of many ancient relics in the Central Java and Yogyakarta (Fig. 2).

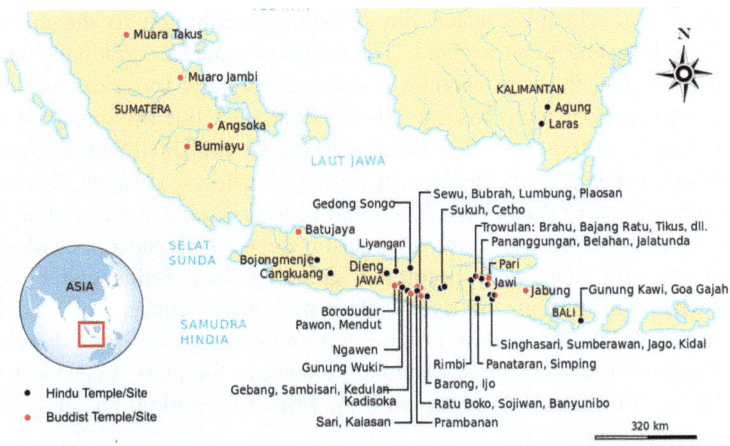

Fig. 2. Distribution of temples and archaeological sites of Java (www.wikipedia.org/wiki/Daftar_candi_di_Indonesia)

4.1.1 Temples on the Foot Slopes of Merapi

In the year 1966 Sambisari temple was found by a farmer. When discovered, it was buried by sediments of about 6 m thick. The sediments covering the temple comprise volcanic materials including sand, tuff, and stones with the grainsize of granule to boulder. As it is displayed on the bank of River Kuning, the deposits show various sedimentary structures such as imbrication and carpet tract, indicating that they are lahar deposit.

At another village namely Kadisoka, there is a small Hindu temple that most of the building still hidden by eruption deposits of Merapi volcano (Fig. 3). In the embankment there are sedimentary structures which indicate that these materials are the interlayering of lahar, fluvial, pyroclastic falls, and glowing clouds deposits. According to Mulyaningsih (2006a, b), around the Kadisoka Temple, there is an evidence of 4 (four) sequences of Merapi deposit that buried the temple, each of which occurred in the different periods. The oldest eruption period happened about 1900 years ago, or in the II century.

Fig. 3. Kadisoka temple (left) and Pustakasala temple (right) (Doc: the authors)

There is another temple called the Kedulan temple, has founded in the IX century. As that of Sambisari temple, it was covered by about 6 meters thick of volcanic materials. It consists of sand, granule, gravel, and boulders with tuff impurities. Based on the soil profile analyzed by Mulyaningsih (2006a, b), the deposits display an interlayering of lahar and pyroclastic fall sedimentation. In a different time and place, when the Islamic University of Indonesia (UII) would built a library building, there also was found a hidden temple which then given the name Pustakasala (Fig. 3). The ruins of the ancient temple were covered by laharic and pyroclastic deposits of Merapi volcano as well.

4.1.2 Sites on the Foot Slope of Sindoro Volcano

On the southwestern slopes of Sindoro volcano, Liyangan site was found by the sand miners (Fig. 4). This site was buried by volcanic deposits as thick as more than 10 m. Based on the findings of wood charcoal, the ash content at the sediment that buried the site, it can be concluded that the Liyangan Site is buried by eruption products of pyroclastic falls and pyroclastic flows.

Fig. 4. Liyangan archaeological site, on the flank of Sindoro volcano (Doc: the authors)

4.2 Geology and Volcanism in the Central Java

The Java island tectonic setting is controlled by the subduction of the India – Australia plate beneath the Eurasia plate forming a series of active volcanoes stand in the middle part of the island (Fig. 5). Despite there are plenty of active volcanoes, due to this tectonic setting, the Java Island is also subjected to earthquake very often. At the present time, the central Java area still affected by several geological disasters such as 2006 Yogyakarta earthquake, and 2010 Merapi eruption. In the study area, presently, volcano that still active is only Merapi, while Merbabu volcano, Sumbing volcano and Sindoro volcano are classified to be in post volcanism stage.

Merapi volcano, is one of the most active volcanoes in Indonesia. When it erupts, the volcano occasionally spews lava, glowing clouds, and pyroclastic materials out. Merapi always offers opportunities as well as threats to the life of the surrounding community. It is able to act as the source of prosperity and the source of the disaster as well. The magma chamber of Merapi is very shallow, it is supplied continuously by the partial melting of the India-Australia ocean plate that constantly moves to the North,

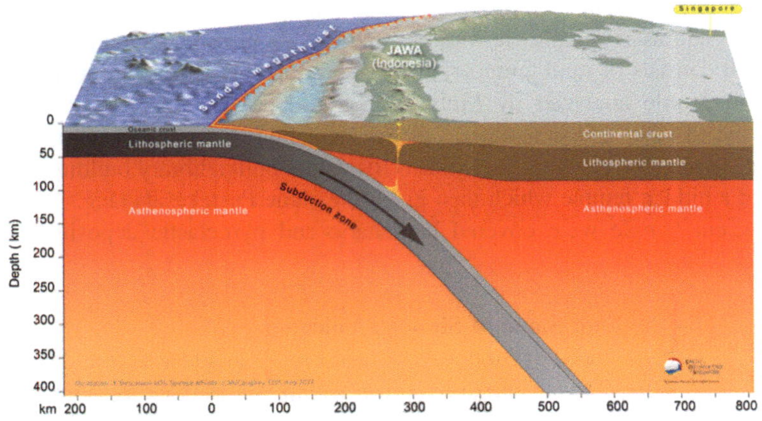

Fig. 5. The tectonic setting of the Java Island, in the convergence of India-Australia plate and Eurasia plate (www.earthobservatory.sg)

subducted under the of the Eurasia continental plate, making the volcano remains alive (Kusumayudha 2013a, b). The distribution of Merapi eruptions deposits are displayed in the geological map of the volcano, as shown in Fig. 6, while situation and condition related to Merapi activities are presented in Figs. 7, 8, 9, and 10.

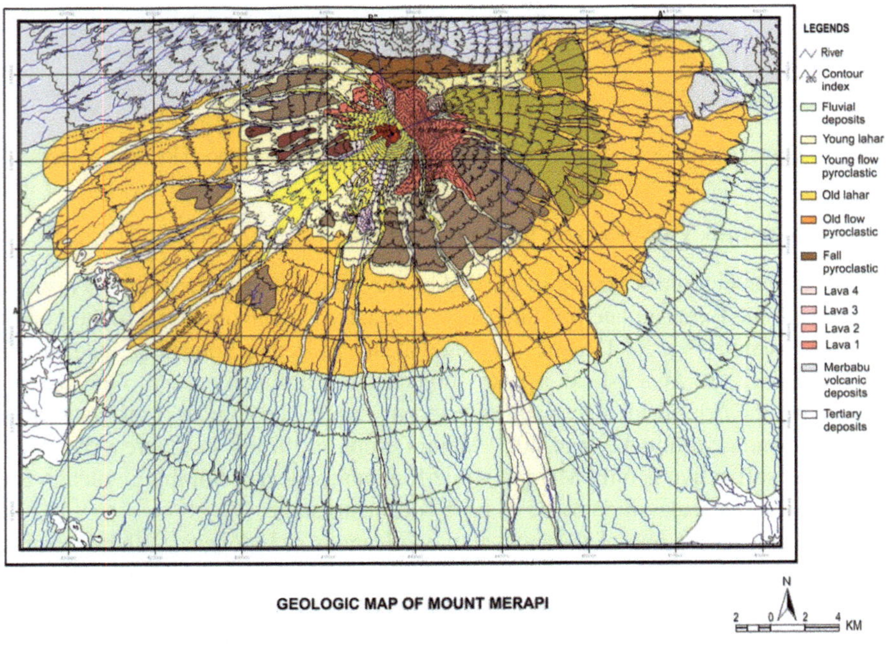

Fig. 6. The geological map of Merapi Volcano and the surrounding area (After vide Kusumayudha et al. 2009)

Fig. 7. Satellite image showing Merapi eruption and lava flow. Acquisition date: November 11, 2010 (courtesy: DigitalGlobe)

Fig. 8. Lahar deposits of 1994 (left) and 2006 (right) (Doc: the authors)

Fig. 9. Kaliadem village buried by pyroclastic deposits, 2010 (Doc: the authors)

Fig. 10. After the eruption of 2010 (left), Borobudur temple covered by volcanic ash (right) (Doc: the Authors)

5 Conclusions

Based on the above discussion, it can be concluded as follows:

1. Some temples in the Central Java and Yogyakarta were discovered in damage condition and/or buried by volcanic materials. They are Sambisari temple, Kadisoka temple, Kedulan temple, Pustakasala temple, and the archeological site of Liyangan.
2. Merapi volcano is one of the most active volcanoes of Indonesia, its eruption in 2006 and 2010 produced pyroclastic flows with a range of about 8 to 17 km distance, burying Kaliadem village of Cangkringan district and surrounding areas, and resulted in areas with a radius of 30 km covered by volcanic ash.
3. Referring to the data that many temples were buried by volcanic materials, it can be concluded that eruptions of volcanoes, especially Merapi volcano had taken part in the destruction of the glory of Mataram Kingdom in the Central Java and Yogyakarta.

References

Kusumadinata, K.: Data Dasar Gunungapi Indonesia, Departemen Pertambangan dan Energi R.I (1979)

Kusumayudha, S.B.: Merapi: A Beautiful Bounty. Penerbit PT Citra Adi Parama, Yogyakarta (2013a)

Kusumayudha, S.B.: Gunungapi Aktif di Indonesia, edisi ke-2. Penerbit PT Citra Aji Parama, Yogyakarta (2013b)

Kusumayudha, S.B., Pratiknyo, P., Riyanto, A.: Hidrokimia airtanah lereng selatan Merapi pasca erupsi 2006. Jurnal Teknologi Mineral **22**(2), 144–153 (2009)

Kusumayudha, S.B.: Merapi: Sebuah Matarantai Sejarah, SKH “Kedaulatan Rakyat”, 22 February 2006 (2006a)

Kusumayudha, S.B., Murwanto, H.: Ancient lake track records, and its influence to modern hydrogeologic condition of the Borobudur National Park Area, Central Java, Indonesia. In: Pros 34th Congress of International Association of Hydrogeologists, Beijing (2006b)

Mulyaningsih, S., Sampurno, S., Zaim, Y., Puradimaja, D.J., McGeehin, J., Bronto, S.: Very old and young temple discoveries in Yogyakarta area: based on Volcano-tratigraphic study. In: Proceeding Volcano International Gathering, Volcano: Life, Prosperity and Harmony, Yogyakarta, 4–10 September 2006, pp. 102–114 (2006a)

Mulyaningsih, S., Sampurno, S., Zaim, Y., Puradimaja, D.J., Bronto, S.: Old building discoveries, parts of lost civilization in Yogyakarta areas: in geologic point of view. In: Proceeding Volcano International Gathering, Volcano: Life, Prosperity and Harmony, Yogyakarta, 4–10 September 2006, pp. 125–133 (2006b)

Poesponegoro, M.D., Notosusanto, N.: Sejarah Nasional Indonesia II, Edisi Pemutakhiran, Penerbit Balai Pustaka, xxii+536 p. (2008)

Van Bemmelen, R.W.: The Geology of Indonesia, Vol. IA, 732 p. Gov. Print. Office, The Hague Martinus Nijhoff (1949)

Zen, M.T.: Enigma Merapi, Sarasehan Merapi dan Sejarah Mataram, UPN “Veteran” Yogyakarta. www.digitalglobe.org, www.earthobservatory.sg, www.wikipedia.org/wiki/Daftar_candi_di_Indonesia (2006)

Causal Analysis and Stability Evaluation of Loess Landslide in Yili Region of Xinjiang - A Case Study of Alar Village Landslide

Fei Ai[1(✉)], Fan Zhou[2], Wanlin Peng[3], Jian Liu[1], Xiuping Yan[3], and Pengfei Chen[3]

[1] State Key Laboratory of Geomechanics and Geotechnical Engineering, Institute of Rock and Soil Mechanics, Chinese Academy of Science, Wuhan, China
aifei_2007@hotmail.com

[2] Hubei Highway Intelligent Maintenance Technology Corporation Limited, Hubei, China

[3] Third Geological Brigade, Geological, Mineral Exploration and Development Bureau of Xinjiang Uygur Autonomous Region, Korla, China

Abstract. Loess landslides are pervasive in the Yili region of Xinjiang province. Loess landslides occur with some inherent conditions and external triggers. Taking the Alar Village landslide in Yili as an example, first, detailed investigation of the inherent conditions of landslides were investigated, such as slope lithologies, geological and hydrogeological conditions. Then, physical and mechanical properties of the landslide materials are measured in laboratory. Experiment results showed that rainfall has a great influence on the physical and mechanical parameters of loess. Under natural conditions, the internal friction angle of loess is 24.5° and the cohesion is 28.4 kPa, while under saturated condition, the mechanical parameters of loess decreased sharply, and the internal friction angle is 16.5°, cohesion of 11.0 kPa. Combined with external triggering factors such as rainfall and earthquake, landslide stability is evaluated by two-dimensional limit equilibrium and finite element method. The simulation results indicated that the factor of safety of slope is the lowest under the earthquake condition. With the groundwater level gradually increased, the factor of safety of slope gradually decreased. The research results can provide theoretical guidance for landslide prevention and control in Yili loess area.

1 Introduction

The loess in the Yili River Valley of Xinjiang is widely distributed and the loess landslide geohazard is the most typical type of geological hazards in the region. The number of loess landslides accounts for 80% of the landslides in Xinjiang (Shi et al. 2018). The study of the formation mechanism and inducing factors of loess landslides is an important theoretical basis to mitigate loess landslides (Ma 2015; Carey et al. 2017; Zhang et al. 2017). A lot of research work have been carried out in view of the formation mechanism, inducing factors, deformation characteristics and mechanical properties of loess on Yili loess landslides.

J. Wasowski and T. Dijkstra (Eds.): GeoMEast 2018, SUCI, pp. 94–104, 2019.
https://doi.org/10.1007/978-3-030-02032-3_9

Zhang et al. (2004) analyzed the formation conditions of loess landslides in Xinyuan Mountain area through geological, topographical and slope stress conditions. Yin et al. (2009) studied the relationship of physical properties and dynamics of Yili loess with landslides. An et al. (2010) analyzed the effects of precipitation and snow melting, earthquakes, vegetation, and human engineering activities on the loess landslide in the Yili region. Wei et al. (2017) and Zhuang et al. (2018) studied the mechanism of the landslide in Piliqing River of Yining and pointed out that the infiltration of ice and snow was the direct cause of landslide. Wang and Mai (2016) studied the formation mechanism and deformation characteristics of the Jialangpute loess landslide while. Luan et al. (2016) performed a three-dimensional stability analysis of the Jialangpute loess landslide. Wang et al. (2016) used the limit equilibrium and finite element strength reduction methods to calculate the landslide stability of loess landslides under rainfall infiltration and earthquake conditions.

In this paper, detailed investigation of the inherent conditions of Alar Village loess landslides were investigated, the physical and mechanical properties of the landslide materials are measured in laboratory. Landslide stability is evaluated by two-dimensional limit equilibrium and finite element method. The research results can provide theoretical guidance for landslide prevention and control in Yili loess area.

2 Geological Conditions

2.1 Location, Climate and Hydraulic Information

The study area is located on the south bank of Mohur river of Alar Village, Gongliu County, Yili region, Xinjiang with the geographical coordinates: longitude 82°42′ 01.38″–82°42′45.81″ and north latitude 43°14′03.15″–43°15′10.27″, as shown in Figs. 1 and 3.

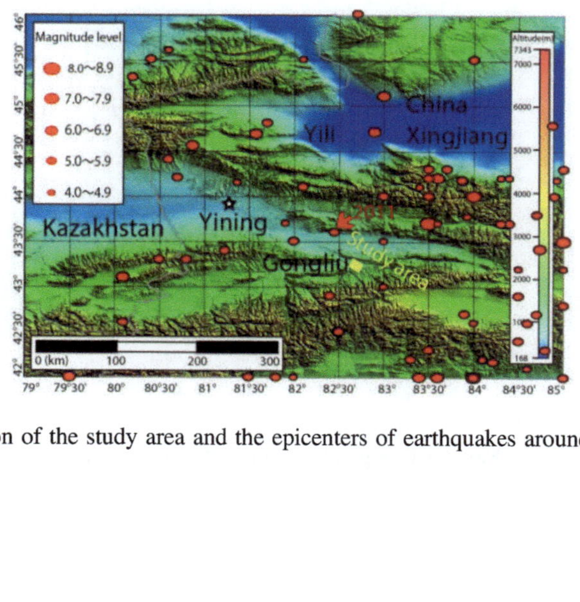

Fig. 1. Location of the study area and the epicenters of earthquakes around the study area

The study area belongs to the continental northern temperate climate. The temperature difference between day and night is large. The summer heat is less and the winter cold is also less, while the spring temperature rises fast and the autumn temperature drops fast. The average annual temperature is around 8.3 °C, as shown in Fig. 2.

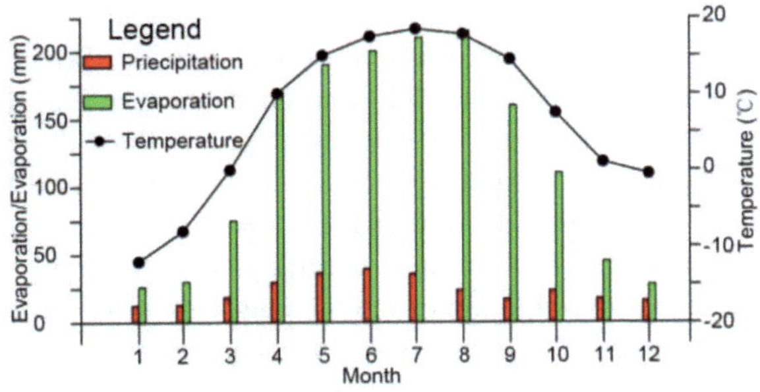

Fig. 2. Precipitation, evaporation and temperature data of the study area

The precipitation in the study area increases in April and reaches the maximum in July. The average annual precipitation is between 300 and 600 mm, and the average annual evaporation is about 1450 mm, as shown in Fig. 2.

The Mohur river in the study area flows from the northern side of the study area to the east-west direction, with an average annual runoff of 59.9 × 10^6 m^3 and a total catchment area of 213 km^2, as shown in Fig. 3.

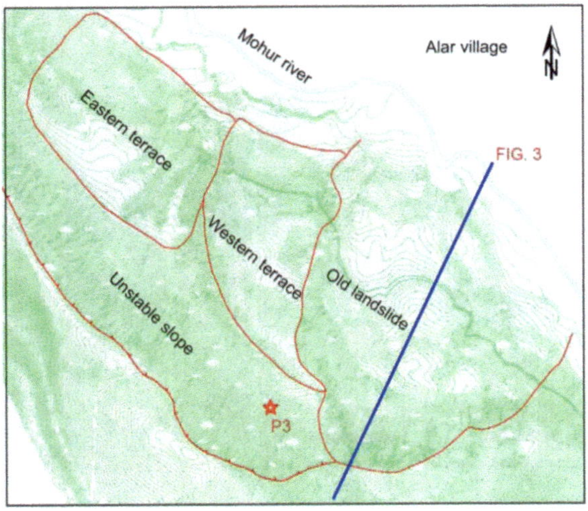

Fig. 3. Topographic map of the study area

2.2 Topography

The geomorphological unit of the study area is a hilly area, as shown in Figs. 3 and 4. The topography of the area is undulating with high in the west and low in the west. The elevation is between 1138–1522 m. The secondary landform types can be divided into southern low mountains, loess platforms, landslide accumulation bodies and gently inclined alluvial valley plains.

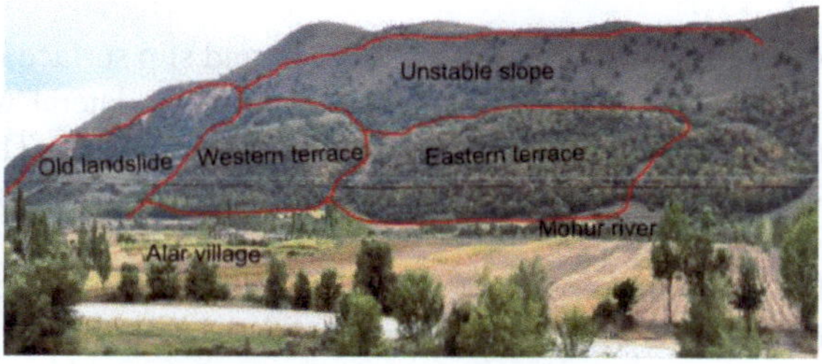

Fig. 4. Field photo of the study area

2.3 Lithology

The lithology of study area includes: the Quaternary Holocene landslide deposits distributed on the old landslide deposits which is composed of silty soil, gravel sand, and fully weathered mudstone mixed gravel layer, the Quaternary upper-Pleistocene aeolian strata which is pale yellow to gray-yellow eolian loess showing a slightly dense-medium-like, pore development, collapsibility, the mid-Pleistocene ice-water sediments with lithology of pebbles and the Neogene brown red-brown mudstone, as shown in Fig. 5.

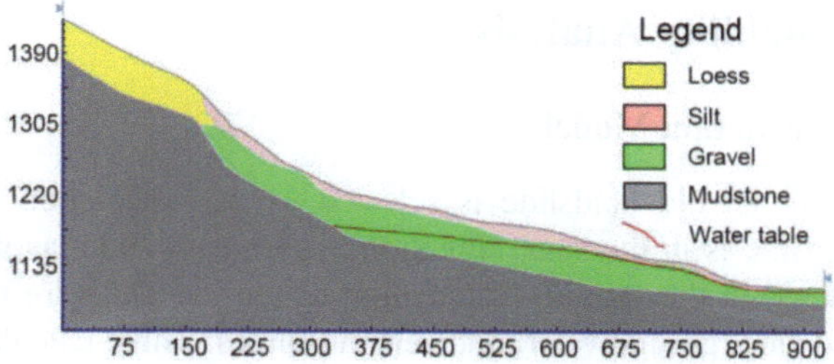

Fig. 5. Profile map of the old landslide

2.4 Earthquake

The study area belongs to the western of the northern Tianshan earthquake zone. The neo-tectonic movement is strong. According to the China earthquake parameters zoning map, the basic seismic intensity in the study area is VIII degrees and the ground motion peak acceleration is 0.2 g. On November 1st, 2011, an earthquake measuring 6.0 on the Richter scale occurred at the junction of Gongliu County and Yining County in Yili. The focal depth was 28 km and the epicenter was about 46 km from the study area, as shown in Fig. 1.

3 Landslide Characteristics and Formation Mechanism

3.1 Spatial Morphology and Basic Characteristics

The morphology of the old landslide likes circle chair and slip surface is circular, with a length of 700 m in the vertical direction, 200–600 m in the lateral width, an area of about 35×10^4 m^2, and a volume of more than 1000×10^4 m^3. The original slope angle of landslide is 30–35°, the length is 700 m, and the width is 200–600 m, as shown in Figs. 3 and 4.

The back wall of the landslide is 60 m high, and the slope angle is 47–50°. The upper part of the back wall is exposed. The trailing edge forms an arc-type tensile crack with 245 m long, which roughly parallel to the back wall of the landslide. The tensile crack formed a shifting surface with a displacement height of 30–50 cm.

3.2 Formation Mechanism

The landslide accumulation body is not formed by a one-time overall sliding but is a traction-multiple sliding complex. The multiple slides mainly slipped along the contact surface between the ice-water sediments and the fully weathered mudstone layer. The slide soil is fully weathered mudstone, and the sliding surface is circular. The loess landslide was judged to be an old landslide with multiple slides at the mudstone-loess interface. The mechanism of old landslide is rotational slide and the unstable slope above old landslide is prone to planar sliding.

4 Landslide Stability Analysis

4.1 Stability Calculation Model

The central axis of the old landslide is selected as the calculation profile, and the direction of the profile is in the same direction as the main slide, as shown in Fig. 3. The loads considered in the stability calculation of the landslide are mainly the body weight of the landslide, groundwater, and seismic forces. Three calculation conditions are considered: natural conditions with sliding body weight, natural + storm condition with sliding body and osmotic pressure, natural + seismic condition with sliding body weight and seismic force. The earthquake acceleration value is taken of 0.20 g. The physical and mechanical tests of mudstone, gravel, silt and loess under natural and saturated conditions were conducted respectively, and the physical and mechanical parameters of rock and soil were obtained as shown in Table 1 and Fig. 6.

Table 1. Physical and geotechnical properties of the old landslide

Lithology	State	Weight (kN/m^3)	Young's modulus (MPa)	Poisson's ratio	Cohesion (kPa)	Friction angle (°)
Loess	Natural	20.5	75	0.39	28.4	24.5
	Saturated	21.5	60	0.43	11.0	16.5
Silt	Natural	16.9	70	0.39	30.4	22.6
	Saturated	17.8	55	0.43	12.2	15.4
Gravel	Natural	20.1	80	0.38	15	25
	Saturated	21.1	64	0.42	13.5	22.5
Mudstone	Natural	25.5	800	0.28	33.5	38.07
	Saturated	25.7	600	0.31	10.0	28.5

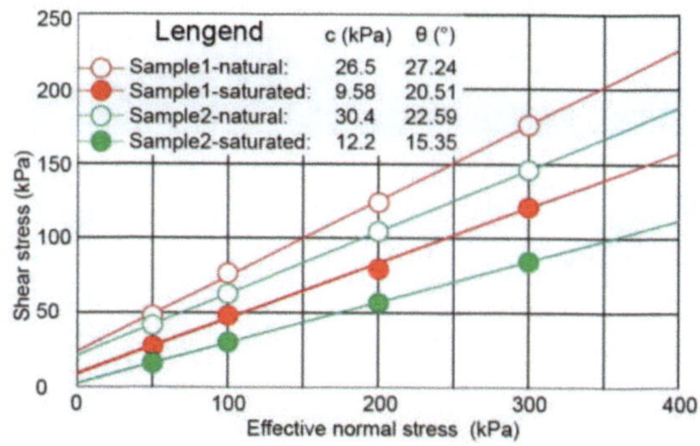

Fig. 6. Direct shear test results of loess in natural and saturated conditions

4.2 Stability Calculation Results

Using two-dimensional limit equilibrium and two-dimensional finite element method to calculate the factor of safety of the landslide in different working conditions, the results are shown in Table 2. The calculation results show that under natural conditions, the landslide factor of safety is slightly greater than 1.74 and is in stable state. When a rainstorm or an earthquake is encountered, the factor of safety is less than 1.1 and is in an unstable state. The slope may be instability and landslides in the snow-melting season, infiltration of rainfall and earthquakes conditions.

Using the finite element method to calculate the horizontal displacement and vertical displacement of the landslide under seismic condition, the results are shown in Fig. 7. The calculation results show that under the earthquake conditions, the maximum horizontal displacement of the landslide is 0.27 m, which occurs in the back wall of old landslide. The maximum vertical displacement of the landslide is 0.15 m, which occurs in the surface part of the back wall of the old landslide.

Table 2. Factor of safety results in different conditions by two methods

Conditions	Factor of safety		Stability statuses
	Two-dimensional limit equilibrium method	Two-dimensional finite difference method	
Natural	1.74	1.76	Stable
Rainfall	1.09	1.13	Basically stable
Earthquake	1.03	1.06	Understable

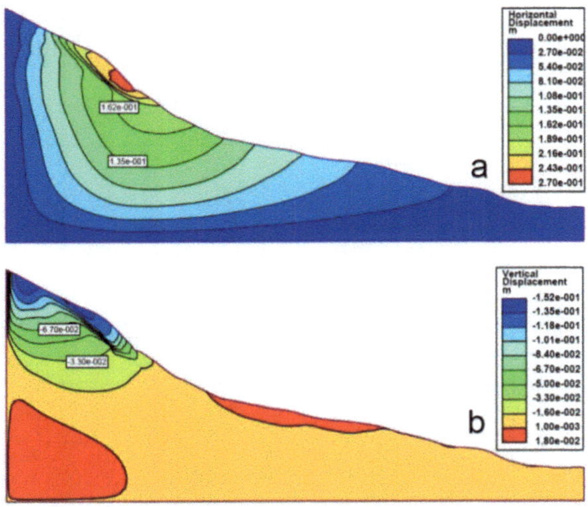

Fig. 7. Contour plots of displacement data of old landslide simulated by two-dimension finite element method: a. horizontal displacement, b. vertical displacement

4.3 Displacement Monitoring Results

Since 2010, a number of monitoring points have been laid at the slope of the trailing edge of the old landslide to measure the horizontal and vertical displacement changes of the slope, as shown in Fig. 3. As of the end of September 2013, the measured values of horizontal displacement and vertical displacement of the P3 monitoring point are shown in Fig. 8. The total horizontal displacement of the P3 point was 0.22 m and the vertical displacement was 0.2 m. Overall, the horizontal displacement and the vertical displacement of the P3 show a gradually increasing trend.

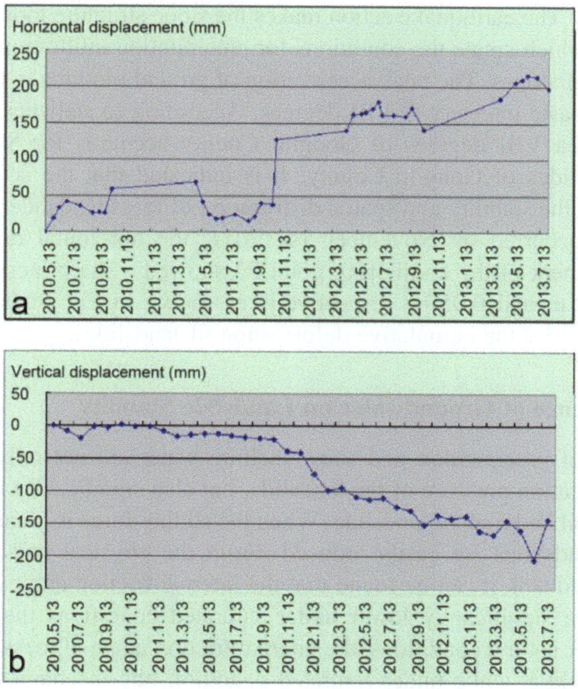

Fig. 8. Horizontal and vertical displacement data of the monitoring point P3: a. horizontal displacement, b. vertical displacement

5 Discussion

5.1 The Influence of Topographic Conditions on Landslide Stability

The study area is located in low mountain with a slope angle of 25–40° and slope orientation consistent with the rock formations. The continuous erosion of the water flow of the Mohur river in the front of the old landslide provides a certain amount of sliding force and air space conditions for the deformation of the landslide. It is not conducive to the stability of landslides and is likely to trigger landslides.

The back wall of the old landslide is 60 m high with a gradient of 47–50°. The upper part of the back wall is exposed and the traces of rain erosion are locally visible. The trailing edges of back wall gradually forms arc-type tensile cracks. Landslides may occur due to instability when the snow melting season, rainfall infiltration and earthquake occur.

5.2 The Influence of Earthquakes on Landslide Stability

Earthquakes are one of important triggering factors of geological hazards. There are two aspects to the inducement of landslides: one is the triggering effect. The earthquake causes the slippage of the slope near the critical stable state. The second is the cumulative effect. The earthquake action makes the slope structure loose and slopes are prone to cracks which create the conditions for precipitation infiltration and promoting unstable failure of slopes. The peak acceleration of ground motions in the study area is 0.20 g, and the basic intensity is VIII degrees. According to statistics, the number of landslides in Area VIII degrees of Gongliu County accounts for 84% of the total number of landslides of Gongliu County. It is indicated that the seismic activity is closely related to the stability and spatial distribution of loess landslide. Affected by the 6.0 earthquake occurred on November 1st, 2011, the horizontal displacement and vertical displacement of the monitoring point P3 showed a sharp increase, which was increased by 0.1 m and 0.05 m, respectively, as shown in Fig. 8. It shows that the earthquake caused by the cumulative deformation of landslides.

5.3 The Influence of Groundwater on Landslide Stability

The infiltration of precipitation and snow melting water are not only increasing the gravity and infiltration pressure of the landslide, but also significantly reduce the shear strength of the soil and rock of landslide. When the sliding force is greater than the anti-sliding force, landslides are easily induced. From the physical and mechanical test results of rock and soil, it can be found that the internal friction angle and cohesion of rock and soil are significantly lower under saturated conditions than under natural conditions. Under natural conditions, the internal friction angle of loess is 24.5° and the cohesion is 28.4 kPa, while under saturated condition, the mechanical parameters of loess decreased sharply, and the internal friction angle is 16.5°, cohesion of 11.0 kPa, as shown in Fig. 6. Compared with natural conditions, the internal friction angle and cohesion of soil under saturated conditions are reduced by 33% and 62%, respectively. The factor of safety of slope under different groundwater conditions is shown in Fig. 9. The calculation results show that the factor of safety of the slope gradually decreases as the groundwater table rises.

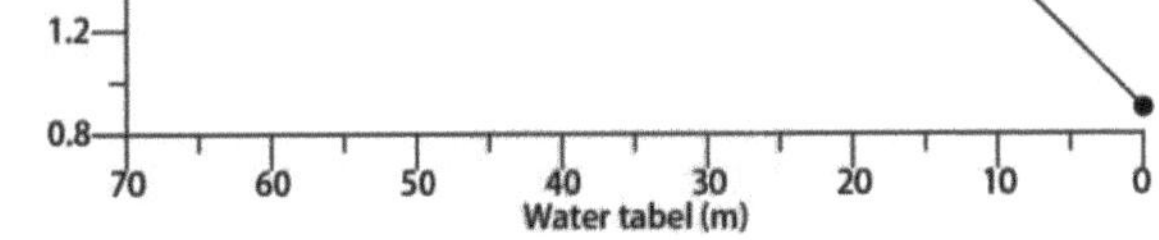

Fig. 9. The relationship between the groundwater table and factor of safety of slope

6 Conclusions

Based on the geological conditions, laboratory test and slope stability analysis, the causal analysis and stability evaluation of loess landslide in Yili region of Xinjiang are analyzed. The conclusions are as follows:

The detailed investigation of the inherent conditions of landslides were investigated, such as slope lithologies, geological results and hydrogeological conditions. It found that the back wall of the old landslide is 60 m high with a gradient of 47–50°. The trailing edges of back wall gradually forms arc-type tensile cracks. Landslides may occur due to instability when the snow melting season, rainfall infiltration and earthquake occur.

The experiment results showed that rainfall has a great influence on the physical and mechanical parameters of loess. Under natural conditions, the internal friction angle of loess is 24.5° and the cohesion is 28.4 kPa, while under saturated condition, the mechanical parameters of loess decreased sharply, and the internal friction angle is 16.5°, cohesion of 11.0 kPa.

Landslide stability is evaluated by two-dimensional limit equilibrium and finite element method. The simulation results indicated that the factor of safety of slope is the lowest under the earthquake condition. With the groundwater level gradually increased, the factor of safety of slope gradually decreased.

Affected by the 6.0 earthquake occurred on November 1st, 2011, the horizontal displacement and vertical displacement of the monitoring point P3 showed a sharp increase, which was increased by 0.1 m and 0.05 m, respectively. The finite element numerical simulation of the slope displacement values is basically consistent with the monitoring displacement values. The research results can provide theoretical guidance for landslide prevention and control in Yili loess area.

Acknowledgments. The financial supports from the National Natural Science Foundation of China (No. 41602320) are gratefully acknowledged. We also like to acknowledge the reviewers for their constructive remarks.

References

An, H., Liu, P.: Genesis and influencing factors of loess landslides in Yili region in Xinjiang. Geol. Hazards Environ. Prot. **21**(3), 22–25 (2010)

Carey, J.M., et al.: Dynamic liquefaction of shear zones in intact loess during simulated earthquake loading. Landslides **14**(3), 789–804 (2017)

Luan, Z., et al.: Analysis on geological features and stability of Jialangpute loess landslide group. J. Shihezi Univ.: Nat. Sci. **34**(5), 643–648 (2016)

Ma, L.: The effect of moisture content on the loess physical properties. Adv. Mater. Res. **1065**, 81–84 (2015)

Shi, J.S., et al.: Neogene clay and its relation to landslides in the southwestern Loess Plateau. China. Environ. Earth Sci. **77**(5), 204 (2018)

Wang, X., Mai, Z.: Sliding mechanism and deformation characteristics of typical large loess landslides in Yili Xinjiang. J. Water Resour. Arch. Eng. **14**(4), 195–200 (2016)

Wang, Y., et al.: Stability analysis and case study of rainfall and disturbance on loess slope. In: International Conference on Smart City and Systems Engineering, pp. 64–67 (2016)
Wei, Y., et al.: Analysis of formation mechanism of Piliqinghe landslide in Yining County, Xinjiang Province. Chin. J. Geol. Hazard Control. **28**(4), 23–26 (2017)
Yin, G., et al.: Physical index, dynamic property and landslide of Yili loess. Arid Geogr. **32**(6), 899–905 (2009)
Zhang, H., et al.: Forming conditions and prevention treatment of loess landslide in the Xinyuan mountain area, Xinjiang. Xinjiang Geol. **22**(3), 233–237 (2004)
Zhang, Z., et al.: Seismic performance of loess-mudstone slope in Tianshui-Centrifuge model tests and numerical analysis. Eng. Geol. **222**, 225–235 (2017)
Zhuang, M., et al.: Type and characteristics of loess landslides in Piliqing River, in Yili of Xinjiang Uygur autonomous region. Chin. J. Geol. Hazard Control. **29**(1), 54–59 (2018)

Artificial Neural Networks for Rock and Soil Cutting Slopes Stability Condition Prediction

Joaquim Tinoco[1,2(✉)], António Gomes Correia[1], Paulo Cortez[2], and David Toll[3]

[1] ISISE – Institute for Sustainability and Innovation in Structural Engineering, School of Engineering, University of Minho, Guimarães, Portugal
jtinoco@civil.uminho.pt
[2] ALGORITMI Research Center/Department of Information Systems, School of Engineering, University of Minho, Guimarães, Portugal
[3] School of Engineering and Computing Sciences, University of Durham, Durham, UK

Abstract. This study aims to develop a tool able to help decision makers to find the best strategy for slopes management tasks. It is known that one of the main challenges nowadays for every developed or countries undergoing development is to keep operational under all conditions their transportations infrastructure. However, due to the network extension and increased budget constraints such challenge is even more difficult to accomplish. Keeping in mind the strong impact of a slope failure in the transportation infrastructure it is important to develop tools able to help minimizing this situation. Accordingly, and in order to achieve this goal, the high flexible learning capabilities of Artificial Neural Networks (ANNs) were applied in the development of a classification tool aiming to identify the stability condition of a rock and soil cutting slopes, keeping in mind the use of information usually collected during routine inspections activities (visual information) to feed them. For that, it was followed a nominal classification strategy and, in order to overcome the problem of imbalanced data, three training sampling approaches were explored: no resampling, SMOTE (Synthetic Minority Over-sampling Technique) and Oversampling. The achieved results are presented and discussed, comparing the achieved performance for both slope types (rock and soil cuttings) as well as the effect of the sampling approaches. An input-sensitivity analysis was applied, allowing to measure the relative influence of each model attribute.

1 Introduction

A nowadays challenge concerning to the transportation network, mainly for every developed or countries undergoing development that have invested and keep investing to build a safe and functional transportation network, is how to keep it operational under all conditions, keeping in mind its extension and the increased budget limitation for maintenance and repair tasks. Thus, taken into account the key importance of the transportation system in modern societies, it is fundamental to develop new tools able to help in its management.

J. Wasowski and T. Dijkstra (Eds.): GeoMEast 2018, SUCI, pp. 105–114, 2019.
https://doi.org/10.1007/978-3-030-02032-3_10

Keeping in mind the strong impact of a slope failure in the transportation infrastructure, namely for the railway it is important to develop tools able to help minimizing this situation.

Over time, several efforts have been made toward the development of a system to detect slope failures. However, most of the systems were developed for natural slopes, presenting some limitations when applied to engineered (human made) slopes. In addition, they have limited applicability as most of them were developed based on particular case studies or using small databases. Furthermore, another aspect that can limit its applicability is related with the information required to feed them, such as data taken from complex tests or from expensive monitoring systems. Pourkhosravani and Kalantari (2011) summarized in their work some of the current methods for slope failure detection, which were grouped into Limit Equilibrium (LE) methods, Numerical Analysis methods, Artificial Neural Networks and Limit Analysis methods. There are also approaches based on finite elements methods (Suchomel et al. 2010), reliability analysis (Sivakumar and Murthy 2005; Husein Malkawi et al. 2000), as well as some methods making use of data mining (DM) algorithms (Cheng and Hoang 2016; Ahangar-Asr et al. 2010; Yao et al. 2008; Kang et al. 2017; Kang and Li 2016; Suman et al. 2016). More recently, a new flexible statistical system was proposed by Pinheiro et al. (2015), based on the assessment of different factors that affect the behavior of a given slope, which are weighted in order to calculate a final indicator of the slope stability condition.

In summary, most of the approaches so far proposed share the main limitations, which are related with its applicability domain or dependency on information that is difficult to obtain. In fact, the assessment of the stability condition of given slope is a multi-variable problem characterized by a high dimensionality.

Artificial Neural Networks (ANNs) are one of the most well known Data Mining (DM) algorithms, which have been applied with success in different knowledge domains, such as web search, spam filters, recommender systems, and fraud detection (Domingos 2012). Also in civil engineering field, several applications can be found. For example, ANNs were applied in the study of physical and mechanical properties of jet grouting columns (Tinoco et al. 2014, 2016). Indeed, the high learning capabilities of this algorithm give it the ability to model complex nonlinear mappings. Thus, in this work the high flexible learning capabilities of Artificial Neural Networks (ANNs) were used to predict the stability condition of rock and soil cutting slopes according to a pre-defined classification scale based on four levels (classes). Moreover, one of the underlying premises of this work is to use only information that can be easily obtained. For that, more than fifty variables related with data collected during routine inspections as well as geometric, geological and geographic data were used to feed the models. With this methodology it is intended to identify critical zones for which more detailed information can then be obtained in order to perform more detailed stability analysis. Such novel approach is intended to support railway network management companies to allocate the available funds to the priority assets according to their stability condition.

2 Methodology

Data characterization. To fit the proposed models for stability condition identification, from this point referred to as EHC (Earthwork Hazard Category (Power et al. 2016)), of rock and soil cutting slopes two database were compiled respectively. Both databases, containing information collected during routine inspections and complemented with geometric, geological and geographic data of each slope, were gathered by Network Rail workers and are concerned with the railway network of the UK. For each slope a class of the EHC system was defined by the Network Rail Engineers based on their experience/algorithms (Power et al. 2016), which will be assumed as a proxy for the real stability condition of the slope for year 2015. The EHC system comprises 4 classes[1] ("A", "B", "C" and "D") where "A" represents a good stability condition and "D" a bad stability condition. In other words, the expected probability of failure is higher for class "D" and lower for class "A".

Both databases contain a significant number of records. The rock slopes database comprises 5945 records, while the soil cutting slopes database is bigger, having 10928 records available. Figure 1 plots the distribution of EHC classes for each database. From their analysis, it is possible to observe a high asymmetric distribution (imbalanced data), particularly for the rock cutting slopes database. In fact, more than 86% of the rock slopes are classified as "A". Although this type of asymmetric distribution, where most of the slopes present a low probability of failure (class "A"), is normal and desirable from the safety point of view and slope network management, it can represent an important challenge for data-driven models learning, as detailed in next section. The proposed models for EHC identification of rock and soil cutting slopes were fed with more than fifty variables normally collected during routine inspections and complemented with geometric, geographic and geological information. To be precise, 65 variables were used in the rock slopes study and 51 variables in soil cutting slopes. Since the number of analyzed variables is high (65/51), just a few examples of the variables used to feed the models are here enumerated: height, slope angle, presence of rock outcrops, animal activity, presence of boulders, ground cover, rock type, dangerous trees, number of root balls, rock strength, etc.

Modeling. In this work, ANNs were trained to predict EHC of rock and soil cutting slopes. ANNs, although not new, they are supported in a strong background and experience. Indeed, they have been applied in the past with high success in different knowledge domains including in civil engineering (Chou et al. 2016; Correia et al. 2013). There are also some examples of ANNs applications in slope stability analysis (Wang et al. 2005; Cheng et al. 2012).

ANNs are learning machines that were initially inspired in functioning of the human brain (Keniget et al. 2001). The information is processed using iteration among several neurons. This technique is capable of modeling complex non-linear mappings and is robust in exploration of data with noise. In this study we adopt the multilayer

[1] The original EHC system comprised 5 levels (A, B, C, D and E) (Power et al. 2016). However, and due to the reduced number of slopes classified as E, classes D and E were combined in one, named as D.

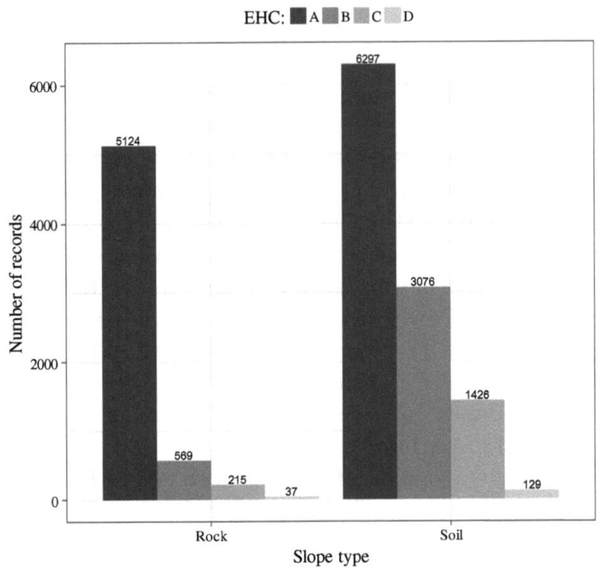

Fig. 1. Rock and soil cutting slopes data distribution by EHC classes

perceptron that contains only feedforward connections, with one hidden layer containing H processing units. Because the ANN performance is sensitive to H (a trade-off between fitting accuracy and generalization capability), a grid search of {0; 2; 4; 6; 8} was adopted under an internal (i.e. applied over training data) three fold cross validation during the learning phase to find the best H value. Under this grid search, the H value that produced the lowest MAE (Mean Absolute Error) was selected, and then the ANN was retrained with all of the training data. The neural function of the hidden nodes was set to the popular logistic function $1/(1 + e^{-x})$.

The problem of EHC determination was approached following a nominal classification strategy, where the network calculates a probability for each EHC level and then is selected the class corresponding to the highest probability.

In addition, and in order to minimize the effect of the imbalanced data (see Fig. 1), Oversampling (Ling and Li 1998) and SMOTE (Synthetic Minority Oversampling Technique) (Chawla et al. 2002) approaches were applied over the training data before fitting the models. When approaching imbalanced classification tasks, where there is at least one target class label with a smaller number of training samples when compared with other target class labels, the simple use of a soft computing training algorithm will lead to data-driven models with better prediction accuracies for the majority classes and worst classification accuracies for the minority classes. Thus, techniques that adjust the training data in order to balance the output class labels, such as Oversampling and SMOTE, are commonly used with imbalanced datasets. In particular, Oversampling is a simple technique that randomly adds samples (with repetition) of the minority classes to the training data, such that the final training set is balanced. SMOTE is a more sophisticated technique that creates "new data" by looking at nearest neighbors to establish a neighborhood and then sampling from within that neighborhood. It operates on the assumptions that the original data is similar because of proximity. We note that the different sampling approaches were applied only to training data, used to fit the data-driven models, and the test data (as provided by the 5-fold procedure) was kept without any change.

For models evaluation and comparison, three classification metrics were adopted: recall, precision and $F_{1\text{-}score}$ (Hastie et al. 2009). The recall measures the ratio of how many cases of a certain class were properly captured by the model. In other words, the recall of a certain class is given by $TruePositives/(TruePositives + FalseNegatives)$. On the other hand, the precision measures the correctness of the model when it predicts a certain class. More specifically, the precision of a certain class is given by $TruePositives/(TruePositives + FalsePositives)$. The $F_{1\text{-}score}$ was also calculated, which represent a trade-off between the recall and precision of a class. The $F_{1\text{-}score}$ correspond to the harmonic mean of precision and recall, according to the following expression:

$$F_{1-score} = 2 \cdot \frac{precision \cdot recall}{precision + recall} \quad (1)$$

For all three metrics, the higher the value, the better are the predictions, ranging from 0% to 100%.

The generalization capacity of the models was accessed through a 5-fold cross-validation approach under 20 runs (Hastie et al. 2009). This means that each modeling setup is trained $5 \times 10 = 100$ times. Also, the three prediction metrics are always computed on test unseen data (as provided by the 5-fold validation procedure).

All experiments were conducted under the R statistical environment (R Team 2009). ANN algorithm was trained using the rminer package (Cortez 2010), which facilitates its implementation, as well as different validation approaches such as the cross validation adopted in this work.

3 Results and Discussion

Following are presented and discussed the achieved performance in EHC determination of both rock and soil cutting slopes based on ANN algorithm, as well as the effect of the three training sampling approaches explored: Normal (no resampling), OVERed (Oversampling) and SMOTEd (SMOTE).

Figure 2 gives an overview of ANNs models performance in rock and soil cutting slopes, based on recall, precision and F1-score.

Concerning to rock cutting slopes study (left side of Fig. 2) the achieved performance is somewhat lower independently of the resampling approach applied. Although a very high performance is observed for class “A” ($F_{1\text{-}score}$ higher than 95%), for class “C” and particularly for class “D”, all models evidence clear difficulties in predicting such classes correctly. In fact, and using $F_{1\text{-}score}$ as reference, the best performance in identification of slopes of class “D” is lower than 14% (see Fig. 2) which was achieved by applying SMOTE resampling approach. From Fig. 3a analysis, which plots the

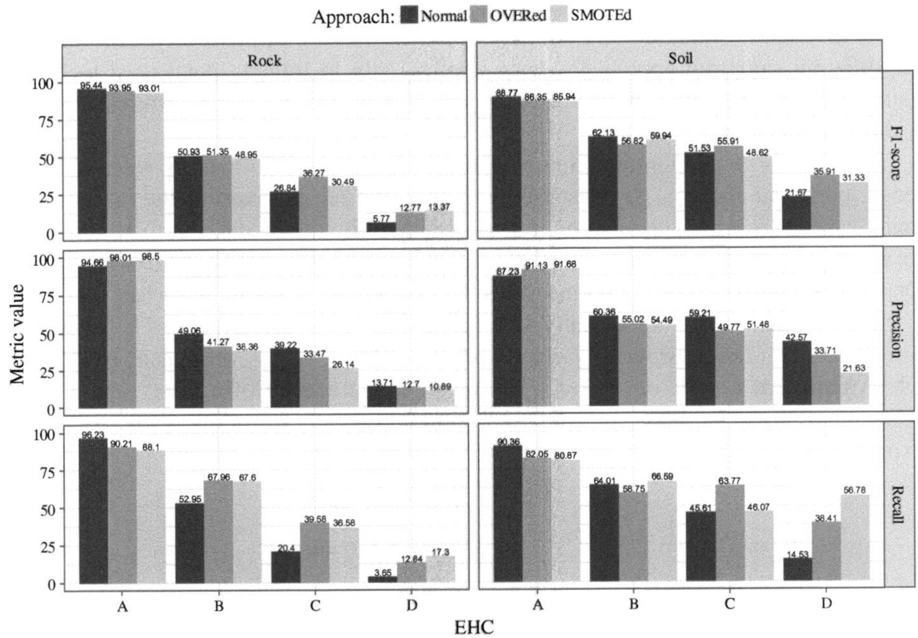

Fig. 2. ANN performance comparison in EHC determination of both rock and soil cutting slopes based on recall, precision and F1-score

relation between observed and predicted EHC values of rock cutting slopes by applying oversampling approach (best overall fit), it is clear the model difficulties in correctly predicting class "C" and particularly class "D", for which the expected probability of failure is higher. As shown, only around 12% of rock cutting slopes classified as "D" are correctly identified, which represents a low performance, namely when compared with soil cutting slopes study as following discussed. Overall, these results show that the methodology applied for EHC determination of rock cutting slopes needs future development in order to overcome this gap.

Relating to soil cutting slopes study, a very promising performance is observed, as shown in Fig. 2 (right side). For example, soil cutting slopes of class "A" can be correctly identified, either with or without resampling. Also for classes "B" and "C" a promising performance is also observed, with an $F_{1\text{-}score}$ around 55%. Concerning the class "D", although an $F_{1\text{-}score}$ lower than 36% was achieved, the obtained value for recall metric around 57% shows a good performance for class "D" identification. Analyzing Fig. 3b that shows the relation between observed and predicted EHC values according to the best overall fit (SMOTEd), one can see that the model performance is very promising. Indeed, when a SMOTE resampling approach is applied, ANN algorithm is able to predict correctly around 57% of soil cutting slopes of class "D", which represents a very promising performance if we take into account that this is the minority class. For class "C", although the accuracy is not so high (around 40%), when not predicted as "C" they are classified as belonging to the closest class, that is, "B" or

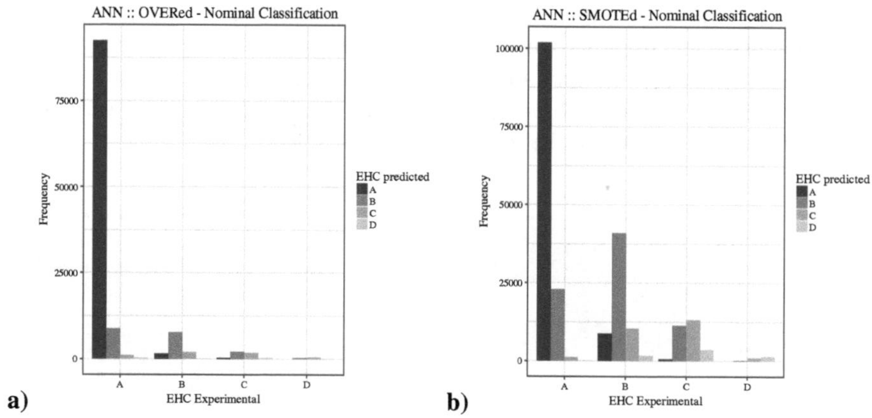

Fig. 3. ANN performance: (a) Rock cutting slopes following an OVERed approach; (b) Soil cutting slopes following a SMOTEd approach

"D". This type of misclassification is also observed for classes "A", "B" and "D", which can be interpreted as positive point. Concerning to classes "A" and "B", the ANN model was able to identify it very accurately.

Comparing the achieved results of rock and soil cutting slopes studies, the proposed models for soil cutting slopes are more effective, namely in the identification of classes "C" and "D" for which the probability of failure is higher (see Fig. 2). A possible explanation for the lower performance, namely for classes "C" and "D" of rock cutting slopes could be related with the EHC classes being assumed as representative of the real stability condition of each slope. Indeed, analyzing the number of slope failures by EHC class for rock slopes there are some indications that the classification attributed to each rock slope could lack of some accuracy as reported in the work of Power et al. (2016), which used the same source of information. It would be expected that most of the failures would occur in slopes of classes "C" and mainly "D". However, for rock slopes such behavior is not observed as reported on Power et al. (2016). In fact, the number of failures for each EHC class is almost constant from classes "A" to "D", particularly when compared with soil cuttings. For example, the number of failures observed in rock cutting slopes of class "C" is only twice higher when compared to class "A". This observation shows that the defined classes for rock slopes have a poor correlation with actual failures.

These results show that a deeper data analysis is required, particularly in the study of rock cutting slopes. For example, the number of variables taken as model attributes might be too high and may be influencing the generalization performance of the models.

Thus, and in order to better understanding the proposed models, the relative importance of each model input was measured based on a sensitive analysis. For that, the methodology proposed by Cortez and Embrechts (2013) was applied. Accordingly, Fig. 4 shows the relative importance of the 20 most relevant variables based on the two models that achieved the overall best performance in EHC determination of rock (left side) and soil (right side) cutting slopes (see Fig. 3). From their analysis, it is observed that the height of the slope is one of the most relevant variables affecting the slope stability condition, either for rock or soil cutting slopes. On the other hand, and particularly for rock slopes, it is observed that the first seven most relevant variables are responsible for around 70% of the total influence.

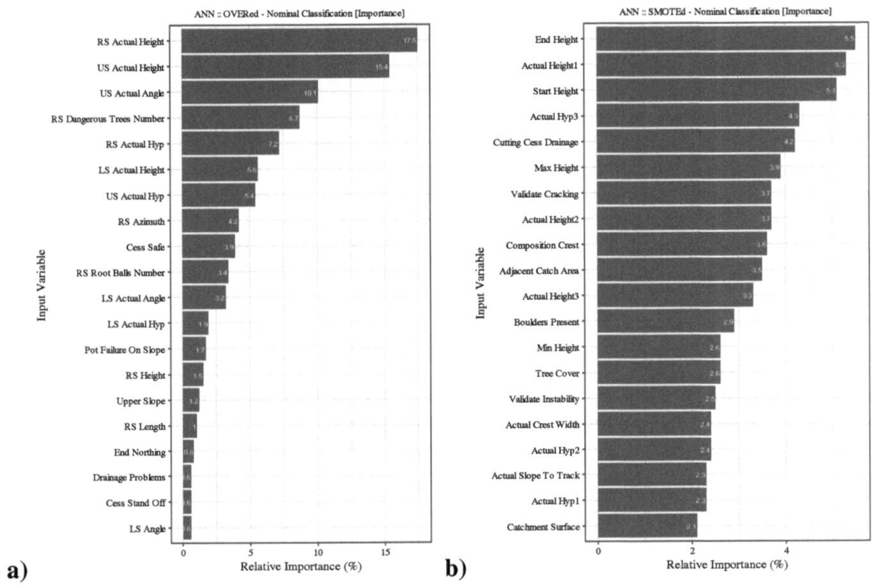

Fig. 4. Relative importance bar plot of the 20 most relevant variables according to ANN models in EHC determination: (a) Rock cutting slopes following an OVERed approach; (b) Soil cutting slopes following a SMOTEd approach

These results, together with achieved performance, suggest a need of revision of the variables used as model attributes. Accordingly, and as a future works, it is intended to apply a more sophisticated feature selection method in order to improve models performance. In particular, it will be explored a multi-objective evolutionary computation method that simultaneously maximizes prediction performance and minimizes the number of inputs used.

As a final observation, and considering the overall performance of all models, it should be underlined the potential of ANNs, or even other soft computing algorithms, in EHC determination of rock and soil cutting slopes.

Acknowledgments. This work was supported by FCT - "Fundação para a Ciência e a Tecnologia", within ISISE, project UID/ECI/04029/2013 as well Project Scope: UID/CEC/00319/2013 and through the post-doctoral Grant fellowship with reference SFRH/BPD/94792/2013. This work was also partly financed by FEDER funds through the Competitivity Factors Operational Programme - COMPETE and by national funds through FCT within the scope of the project POCI-01-0145-FEDER-007633. This work has been also supported by COMPETE: POCI-01-0145-FEDER-007043. A special thanks goes to Network Rail that kindly made available the data (basic earthworks examination data and the Earthworks Hazard Condition scores) used in this work.

References

Ahangar-Asr, A., Faramarzi, A., Javadi, A.: A new approach for prediction of the stability of soil and rock slopes. Eng. Comput. **27**(7), 878–893 (2010)

Chawla, N., Bowyer, K., Hall, L., Kegelmeyer, W.: Smote: synthetic minority oversampling technique. J. Artif. Intell. Res. **32**, 321–357 (2002)

Cheng, M., Hoang, N.: Slope collapse prediction using bayesian framework with k-nearest neighbor density estimation: case study in Taiwan. J. Comput. Civ. Eng. **30**(1), 04014116 (2016)

Cheng, M., Roy, A., Chen, K.: Evolutionary risk preference inference model using fuzzy support vector machine for road slope collapse prediction. Expert Syst. Appl. **39**(2), 1737–1746 (2012)

Chou, J., Yang, K., Lin, J.: Peak shear strength of discrete fiber-reinforced soils computed by machine learning and metaensemble methods. J. Comput. Civ. Eng. 04016036 (2016)

Cortez, P.: Data mining with neural networks and support vector machines using the R/rminer tool. In: Perner, P. (ed.) Advances in Data Mining: Applications and Theoretical Aspects. ICDM 2010. LNAI, vol. 6171, pp. 572–583. Springer (2010)

Cortez, P., Embrechts, M.: Using sensitivity analysis and visualization techniques to open black box data mining models. Inf. Sci. **225**, 1–17 (2013)

Domingos, P.: A few useful things to know about machine learning. Commun. ACM **55**(10), 78–87 (2012)

Correia, A.G., Cortez, P., Tinoco, J., Marques, R.: Artificial intelligence applications in transportation geotechnics. Geotech. Geol. Eng. **31**(3), 861–879 (2013)

Hastie, T., Tibshirani, R., Friedman, J.: The Elements of Statistical Learning: Data Mining, Inference, and Prediction, 2nd edn. Springer-Verlag, New York (2009)

Husein Malkawi, A., Hassan, W., Abdulla, F.: Uncertainty and reliability analysis applied to slope stability. Struct. Saf. **22**(2), 161–187 (2000)

Kang, F., Li, J.: Artificial bee colony algorithm optimized support vector regression for system reliability analysis of slopes. J. Comput. Civ. Eng. 04015040 (2016)

Kang, F., Li, J.S., Wang, Y., Li, J.: Extreme learning machine based surrogate model for analyzing system reliability of soil slopes. Eur. J. Environ. Civ. Eng. **21**(11), 1341–1362 (2017)

Kenig, S., Ben-David, A., Omer, M., Sadeh, A.: Control of properties in injection molding by neural networks. Eng. Appl. Artif. Intell. **14**(6), 819–823 (2001)

Ling, C., Li, C.: Data mining for direct marketing: problems and solutions. In: KDD **98**, pp. 73–79 (1998)

Pinheiro, M., Sanches, S., Miranda, T., Neves, A., Tinoco, J., Ferreira, A., Correia, A.G.: A new empirical system for rock slope stability analysis in exploitation stage. Int. J. Rock Mech. Min. Sci. **76**, 182–191 (2015)

Pourkhosravani, A., Kalantari, B.: A review of current methods for slope stability evaluation. Electron. J. Geotech. Eng. **16** (2011)

Power, C., Mian, J., Spink, T., Abbott, S., Edwards, M.: Development of an evidence-based geotechnical asset management policy for network rail, Great Britain. Procedia Eng. **143**, 726–733 (2016)

Sivakumar Babu, G., Murthy, D.: Reliability analysis of unsaturated soil slopes. J. Geotech. Geoenviron. Eng. **131**, 1423–1428 (2005)

Suchomel, R., et al.: Comparison of different probabilistic methods for predicting stability of a slope in spatially variable c–φ soil. Comput. Geotech. **37**(1), 132–140 (2010)

Suman, S., Khan, S., Das, S., Chand, S.: Slope stability analysis using artificial intelligence techniques. Nat. Hazards **84**(2), 727–748 (2016)

Team, R.: R: A language and environment for statistical computing. R Foundation for Statistical Computing, Vienna, Austria (2009). http://www.r-project.org/

Tinoco, J., Correia, A.G., Cortez, P.: Support vector machines applied to uniaxial compressive strength prediction of jet grouting columns. Comput. Geotech. **55**, 132–140 (2014)

Tinoco, J., Correia, A.G., Cortez, P.: Jet grouting column diameter prediction based on a datadriven approach. Eur. J. Environ. Civ. Eng. 1–22 (2016)

Wang, H., Xu, W., Xu, R.: Slope stability evaluation using back propagation neural networks. Eng. Geol. **80**(3), 302–315 (2005)

Yao, X., Tham, L., Dai, F.: Landslide susceptibility mapping based on support vector machine: a case study on natural slopes of Hong Kong, China. Geomorphology **101**(4), 572–582 (2008)

On the Development of Ground-Based and Drone-Borne Radar System

Tomonori Deguchi(✉), Tomoyuki Sugiyama, and Munemaru Kishimoto

Nittetsu Mining Consultants Co., Ltd., Tokyo, Japan
deguchi@nmconsults.co.jp

Abstract. The interferometric SAR (InSAR) technique is an established analytical tool designed to analyze phase information provided by satellite-mounted synthetic aperture radar (SAR) to provide precise measurements of target displacement to a centimeter level of precision. The InSAR technique has already reached the practical stage in the measurement of ground subsidence in urban areas and in crustal deformation analysis associated with earthquakes and volcanic activities. The authors are currently developing radar systems for on-ground installation and mounting on drones. InSAR analysis can also be applied to the data acquired by these systems. We ultimately aim to develop technologies that will allow us to monitor the stability of opencast mine slopes, landslide slopes, and aging infrastructures.

Measurements taken by ground-based radar systems can be conducted at all times automatically, without human operators. Measurement data can be accumulated at intervals of several minutes to several tens of minutes. Millimeter wave radar is internationally recognized as a collision prevention technology for automobiles. Our group has used this technology to develop a millimeter wave radar system capable of transmitting radio waves at a bandwidth (76 to 77 GHz) approved by the regulatory authorities. In field tests, this radar was verified to detect targets remotely from a distance of about 1.6 km and to measure displacement with an accuracy of about 0.2 mm. In a long-term demonstration test conducted inside an open-pit mine, the radar-collected data clearly captured the precursor phenomena of the slope failure. The radar also successfully detected anomalies corresponding to consolidation subsidence of sediment deposited by slope failure.

The drone-borne radar we are developing will be used for periodical observations to be carried out several times a year. We designed the system to transmit the microwave band (12.8 to 12.95 GHz) over a longer wavelength than the millimeter wave band, as the phase information used for InSAR analysis becomes unstable with time. The frequencies of the transmitted radio waves adopted in this study, including those transmitted by the ground-based radar, were selected to comply with the Radio Law of Japan. The ground-based radar has reached a stage where it can be deployed practically in the field. Real-life use of the drone-borne radar has yet to be realized, though a prototype version of the system is now complete. We plan to develop algorithms for synthetic aperture processing and interferometric analysis in the future.

J. Wasowski and T. Dijkstra (Eds.): GeoMEast 2018, SUCI, pp. 115–122, 2019.
https://doi.org/10.1007/978-3-030-02032-3_11

1 Introduction

Our group is developing two radar systems, a ground-based type and drone-borne type. By applying the InSAR technique to the data measured with these systems, we aim to develop technologies for the monitoring of opencast mine slopes, landslide slopes, and aging infrastructure. In this paper we report the current status of device development and the results of field examinations to evaluate the performance of the systems.

2 Development of Ground-Based Radar System

Our ground-based radar system performs distance measurement by the conventional FM-CW method. The specification is shown in Table 1. The radar system for this device employs a real aperture and rotates in both the horizontal and vertical directions (Fig. 1). The system can acquire two-dimensional images while repeatedly transmitting and receiving millimeter waves when rotated on a rotation table installed on a base such as a tripod. Image resolution is weakened in long-distance measurements by this real aperture radar, as the 3 dB beam width of the antenna coincides with the azimuth resolution. There is little concern, however, that the temporary vibrations issued by the rotations of the stepping motor will hinder measurement, as no synthetic aperture processing is required. This latter feature makes it possible to easily establish a highly sustainable measurement system. The gain of the Cassegrain antenna is determined by the antenna diameter and wavelength. With a millimeter wave system, a high gain can be easily secured with only a small antenna diameter. This feature enables long-distance measurement with lower power consumption.

Table 1. Specification of the ground-based radar system using millimeter radio wave.

Frequency	76.5 GHz
Wavelength	3.92 mm
Band width	1 GHz (adjustable)
Modulation	FM-CW
Radar system	Real aperture
Antenna type	Cassegrain (20 cm)
Antenna power	10 mW (10 dBm)
Antenna gain	40 dBi
Range resolution	0.30 m
Azimuth resolution	1.23 degree
A/D converter	16 bits
Weight	10 kg
Accuracy of displacement	less than 0.2 mm

The images in Fig. 2 show the details of a demonstration test performed on a slope formed in an opencast mine. Figure 2(a) shows the landscape observed from the site of

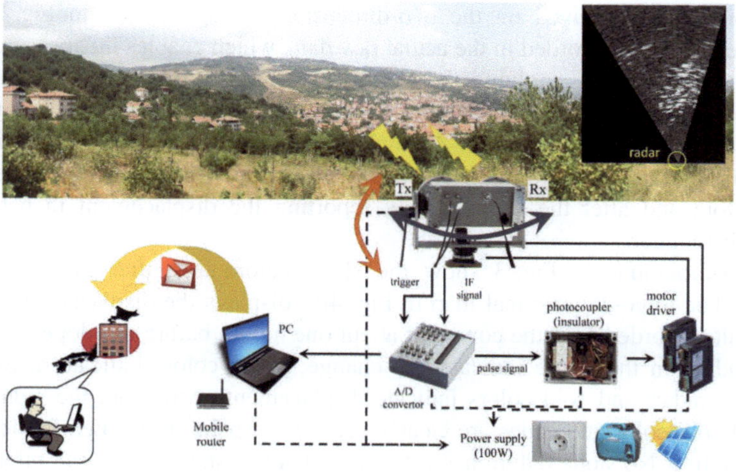

Fig. 1. Conceptual diagram of the ground-based radar system developed.

radar installation, with radio waves transmitted to cover the area framed by the broken red line. Figures 2(b) and (c) are a high-resolution photograph acquired from Google Earth and a power radar image obtained by cutting out the same range, respectively. This demonstration test clearly confirmed that the system could detect a rock slope 1.58 km away from the radar.

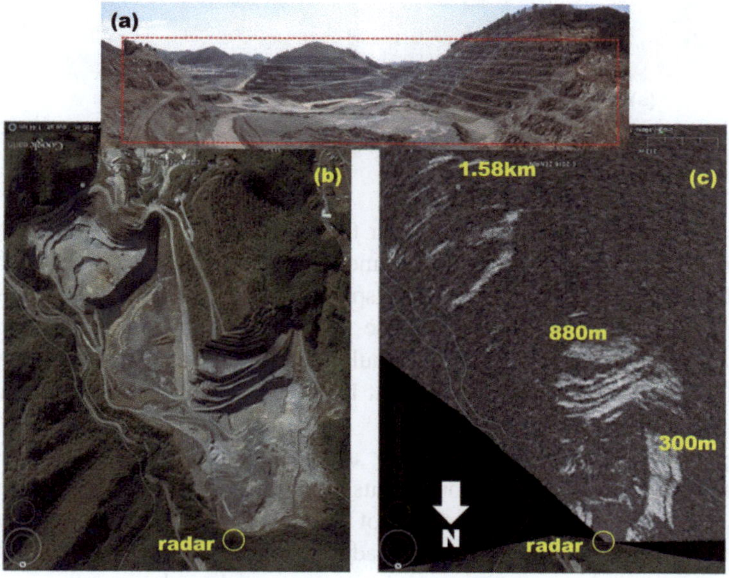

Fig. 2. An example of a two-dimensional radar image acquired of the opencast mine.

Although not displayed on the two-dimensional image shown in Fig. 2(c), the phase information is recorded in the actual raw data, which enables the measurement of minute displacement by applying the InSAR technique. We installed our millimeter wave radar device inside the open pit mine and conducted a long-term demonstration test over the period of 7 months. A small-scale slope failure occurred accidentally within the observation range in the course of the test. The system recorded the ground levels before and after the slope failure, reporting the displacement in between as reported hereunder.

The photographs in Fig. 3 show the slope before and after the slope failure occurred. The three-dimensional map in Fig. 4(a) displays the displacement measurement results recorded over the course of about one month before the slope failure. The green shading in the figure indicates no change; warm colors indicate displacement toward the radar; and cold colors indicate displacement away from the radar. Warm color and cold color anomalies are clearly detected in pairs in the area framed by the solid black line. No slope failure has yet occurred at this stage, so we can infer that the precursor phenomena was captured. The maximum displacement rate measured before slope failure was approximately 4 mm/day.

Fig. 3. Pictures before and after slope failure.

From this measurement result we infer that as the surface layer gradually moved toward the bottom of the slope, the volume at the lower part increased, which was detected by the system anomalously as displacement toward the radar. Likewise, the system anomalously detected the decrease in the volume of the upper part as displacement away from the radar. The cumulative displacement of the precursor phenomena exceeded 30 mm, though it might have been undetectable by the naked eye without careful observation.

A displacement amount that occurring as a precursor phenomenon is difficult to verify quantitatively. The verification data itself is impossible to acquire on the ground, as nobody can know the location and time of a slope failure in advance. Conversely, we can conclude that only a millimeter wave radar device capable of visualizing the subtle surface displacement in space-time coordinates can reach such a result.

Figure 4(b) shows the results of displacement measurements taken for about 2 weeks after the occurrence of the slope failure due to the landslides. We can also detect

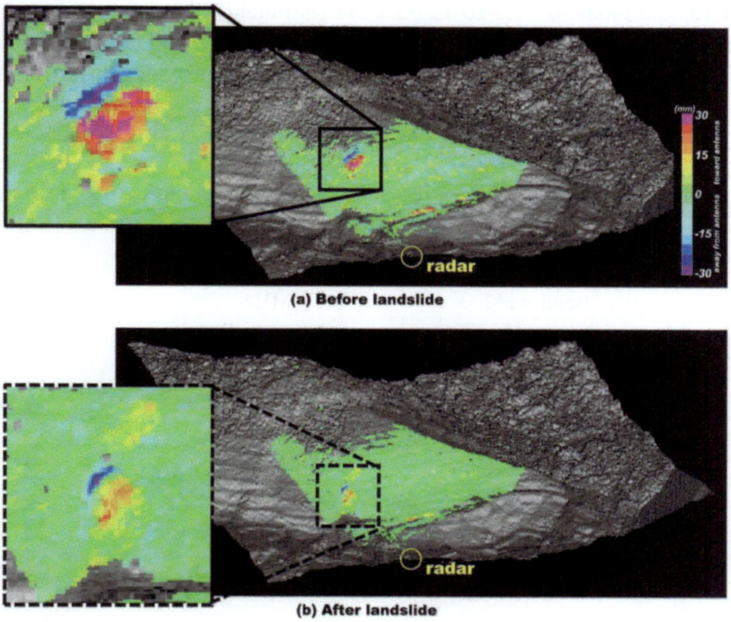

Fig. 4. Three-dimensional displacement map before and after slope failure.

an anomaly in which warm and cold colors are paired, in the area framed by the black broken line. This anomaly falls somewhat below the place where the slope failure occurred. The accumulated sediment is most likely displaced in the subsiding direction with time in the initial stage immediately following the slope failure. We can therefore infer that the consolidation subsidence of the accumulated sediment is measured.

3 Development of Drone-Borne Radar System

Our group is developing a set of technologies to enable the monitoring of slope terrains such as landslide and opencast mine slopes by applying synthetic aperture processing and InSAR analysis to data measured by a drone-mounted radar. Placement on the drone easily secures a line-of-sight vector from multiple directions, since the antenna moves freely. If the orbit of the platform can be flexibly designed, the radar can precisely grasp the three-dimensional displacement of ground surfaces of complicated terrains on its own.

High-precision information on the position of the platform is indispensable for applying synthetic aperture processing. In this research we employed a high-precision GNSS module for drones, a device that realizes centimeter-class positioning using RTK technology. The synthetic aperture processing will be performed by interlocking the raw radar data with the high-accuracy positioning information. We set the radio wave frequency for the radar to 12.8 to 12.95 GHz (Ku-band), the values prescribed in the provisions on "Specific Experimental Testing Authority" under the Japanese Radio Law (the antenna power is 1 W or less in EIRP). Table 2 provides the specifications of the radar.

Table 2. Specification of the drone-borne radar system using microwave.

Frequency	12.9 GHz
Wavelength	23.2 mm
Band width	150 MHz
Modulation	FM-CW
Radar system	Synthetic aperture
Antenna power	15 dBm (31.6 mW)
Antenna gain	15 dBi
Range resolution	1 m
Weight (without battery)	600 g
Power consumption	5.5 Wh

A heavier mounted payload pushes up the power consumption, which shortens the flight time of the drone. Hence, a radar with the lowest possible weight is preferred. In this research we reduced the total weight to 600 g (excluding the power supply) by arranging all of the RF circuits on a single board. Figure 5 shows the appearance of the drone-borne radar device. The system consumes about 5.5 W of power and can operate for approximately 30 min from the power of four AA batteries. Practical tests so far conducted have verified detection up to a distance of about 1.5 km. In the future we will begin developing algorithms for synthetic aperture processing.

4 Conclusions

Our group has developed a ground-based radar system with the aim of constructing a technology to monitor minute displacements of mining slopes and infrastructure buildings by remote sensing techniques. The verification test described in this paper, along with a number of other tests repeatedly performed, have confirmed that high-precision minute deformation measurements can be continued over an extended period of time. Most persuasively of all, our system has clearly visualized the precursor phenomena of an actual slope failure and the behavior of the ground surface presumed as consolidation subsidence of sediment in space-time coordinates.

In future work we will consider a method of correcting the phase delay due to water vapor. We also plan to develop a function to issue slope failure alerts in advance by applying techniques to detect tertiary creep. This may make it possible to update to a system that can objectively evaluate the risk of slope failure.

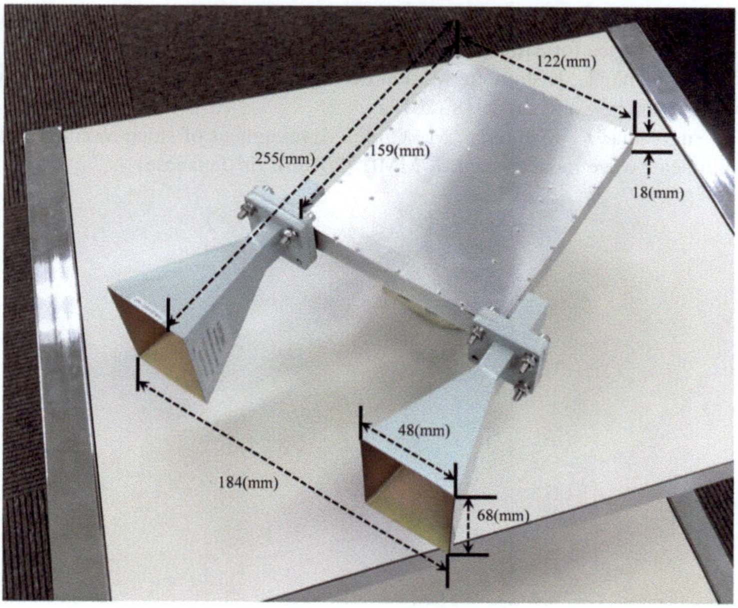

Fig. 5. Appearance and scale of the drone-borne radar.

Drone-borne radar systems have already been developed for commercial use for collision prevention purposes. No precedence has extended to InSAR measurement, however, so the novelty of such an approach is high. Though still in the development stages, we hope to develop algorithms for synthetic aperture processing and InSAR analysis and to advance to demonstration tests towards the practical stage.

Acknowledgments. The development and verification tests for the ground-based radar system have been supported by the members of Nittetsu Mining Co., Ltd. Our work to date on the drone-borne radar has been supported by "Coordinate Funds for Promoting AeroSpace Utilization" from the Ministry of Education, Culture, Sports, Science and Technology (MEXT), JAPAN.

References

Werner, C., Strozzi, T., Wiesmann, A., Wegmuller, U.: Gamma's portable radar interferometer. In: 13th FIG Symposium on Deformation Measurement and Analysis (2008)

Amelung, F., Galloway, D.L., Bell, J.W., Zebker, H.A., Laczniak, R.J.: Sensing the ups and downs of Las Vegas: InSAR reveals structural control of land subsidence and aquifer-system deformation. Geology **27**(6), 483–486 (1999)

Rose, N.D., Hungr, O.: Forecasting potential rock slope failure in open pit mines using the inverse-velocity method. Int. J. Rock Mech. Min. Sci. **44**, 308–320 (2007)

Ogawa, S., Fukunaga, T., Yamagishi, S., Yamada, M., Inaba, T.: 76-GHz high-resolution radar for automatic driving support. SEI Tech. Rev. **192** (2018)

Deguchi, T.: Deformation monitoring using space-borne and ground based radar interferometry. NexGen Technologies for Mining and Fuel Industries, vol. I (2017)

Deguchi, T., Sugiyama, T., Kishimoto, M.: On the R&D project of drone-borne interferometric SAR. World Mining Congress 2018 (2018a)

Deguchi, T., Sugiyama, T., Kishimoto, M.: On the development of slope monitoring technique using millimeter wave radar system. J. MMIJ (2018b). (in Japanese)

Geotechnical Engineering Behavior of Mudstone Formations of Al Wadi Al Gadid Region in Egypt

Mostaf A. Yousef(✉) and Ahmed T. M. Farid

Housing and Building National Research Center, HBRC, Cairo, Egypt
mostafa.abdelfadil@yahoo.com, atfarid2013@gmail.com

Abstract. More than ninety five percent of Egypt land consists of desert regions. According to the large of the desert areas, new cities are constructed at these desert locations. Most of the desert regions are composed of problematic soil formations. Special geotechnical investigation studies must be carefully performed for these problematic soil formations. Al Wadi Al Gadid region which located in the central part of the wesern desert of Egypt is one of the regions which composed of a problematic soil formation. This region has a low population which needs more cities in the future. In this research, the geotechnical engineering behavior of Al Wadi Al Gadid mudstone formations is studied. Different Laboratory tests are performed on different undisturbed samples as bulk samples collected from different locations in the region to predict its geotechnical engineering parameters. Mudstone formation of the area was founded to have a highly swelling behavior under wetting conditions compared to its dry conditions. According to that, highly special geotechnical studies are given to this type of formations to predict its behavior for the different structures established on its formations. Recommendations are given to geotechnical engineers for highly soil investigations on its mudstone formations to predict its characteristics and overcome its change of volume in the future for their projects as water reached the soil under foundation level.

1 Introduction

The term mudrock is the name for claystones, mudstones, siltstones, or shales. These mudrocks refers to the fine-grained, siliciclastic sedimentary rocks in which more than 50% of the particles are smaller than 0.06 mm in size (Blatt et al. 1980, Dick and Shakoor 1992; Grainger 1984). Mudrocks or shales constitute about 45% to 55% of sedimentary rock sequences; thus they are often encountered in engineering construction. However, it is often difficult to obtain good quality, undisturbed samples of mudrocks for laboratory testing because they are weak and sensitive to changes in moisture, drilling pressure, and time (Fam et al. 2003). Therefore, information about its behavior for most engineering projects involving mudrocks is either scarce or assumed. One way to overcome this problem is to develop a methodology that could be used to predict the change of the behavior of mudrock or shale under wetting conditions and reduction of their shear strength and engineering characteristics. researches has been done to relate durability, free swell and swelling behaviors of mudrock or shale to their

J. Wasowski and T. Dijkstra (Eds.): GeoMEast 2018, SUCI, pp. 123–134, 2019.
https://doi.org/10.1007/978-3-030-02032-3_13

lithological characteristics and engineering properties (Dick 1994; Sarman et al. 1994; Greene 2001) and still more work are needed for investigation of their behavior, especially for swelling or shrinkage conditions. Several researchers also studied the soil swelling potential (Holtz and Gibbs 1956; Mohamed 1978; Sridharan and Prakash 2000(where, they define the significant physical characteristics which is effect on the swell behavior. Study of the geometric pattern of the swelling mudstone may eliminate part of the difficulty in adequately predicting its swelling behavior as investigated by (Johnson et al 1973; Sobolevsky and Morgachiev 1973; and Radwan et al. 1989).

In this paper, the study of Wadi Al Gadid region which is located at the centeral part of the western desert of Egypt, where mudstone or shale formation is appeared from surface which inter-bedded by sandstone layers in some places. This mudstone formation of the region is varied in its clay percentages and its composition which causes the effect of the swelling behavior. The swelling heave expected under structures established above this type of mudrock or shale formations should be predicted according to each structure loads. Thus, geotechnical investigations of these sites should be studied carefully for the different location in the region as indicated by (Dhowian 1981; Dhowian 1984). Due to climatic conditions characterized by rare rainfall, these mudstone formations are practically in dry conditions. Cracks can be appeared in structures established directly on these formations without ant treatments, especially for light structures due to the heave of soil as water reached in the future. Consequently, structural problems in buildings have become more frequent and serious if no precautions and soil treatment recommended for these type of problematic mudstone formations in the geotechnical reports.

2 Geology of Study Area

The study area of Al Wadi Al-Gadid located within the centeral part of the western desert of Egypt. Figures 1 and 2 show the location of the study area with respect to Egypt map and with respect to Al Wadi Al Gadid (New Valley) region, respectively. The geology map of the study area has been subject to many investigations and study under the (New Valley Project) by many researchers such as: (Issawi and Jux 1982; Said and Kerdany 1961; Kostandi 1963; and Ghobrial 1967). The Jurassic-Campanian sequence includes the predominantly continental sandstone and clay beds that were formely lumped under the term (Nubia). Recently, upon the detection of the marginal marine shale or mudrock strata which occur presently in outcrops at certain stratigraphical intervals with the sandstone strata below. The proposed study area has different buildings categories ranged from two floors up to six floors which represent light to medium stress on soil. More than 32 boreholes of depths ranged between 10 m and 15 m are drilled in the study area for constructing of residential buildings, warehouses, and roads. Seven additional test pits are executed down from the foundation surface which is three meters to extra four meters below that depth. Undisturbed dry Mudrock samples are extracted from these test pits for laboratory testing.

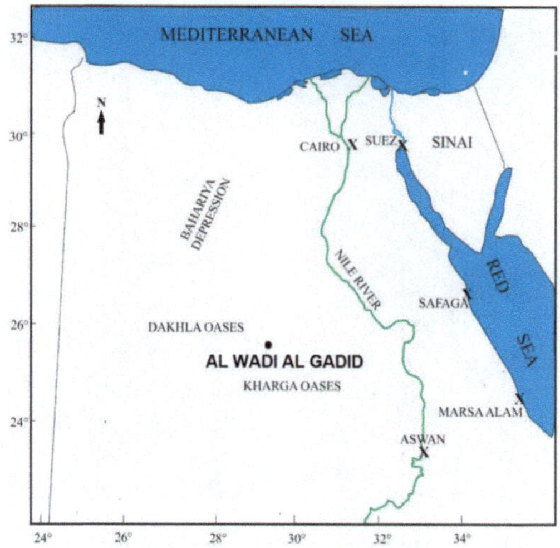

Fig. 1. Location of Al Wadi Al Gadid region on Egypt map

Fig. 2. Location of study area of Al Wadi Al Gadid region on map

Soil investigation performed at the study area shows that, the site consists of medium severely weathered and crushed to closely fractured mudstone formations which appeared from the ground surface down to the end of boring depth. Figure 3 shows sketches for the site mudrock formation at the site surface and also along the side excavation which shows the reddish brown mudstone color and its structural shape.

Fig. 3. Mudstone or shale shape at the site

3 Laboratory Test Program

As the site consists of mudrocks from the surface strata down to end of drilling, special laboratory tests should be performed on this type of formation. Mudrocks of the site consists of mudstone and claystone which could have swell behavior according to their clay content. Tests are performed on different samples collected from the site. In this study, mudrock behavior was investigated by different tests performed on mudrock samples. Swelling tests according to (ASTM D-4546) is performed on mudrock samples to predict the swelling potential and free swell ratios. Atterberg limits are performed on the tested samples to get the activity values of the mudrock samples. Also, hydrometer tests are performed on mudrock samples to get the percentages of clay content samples for classification according to (Potter et al. 1980). X-ray diffraction tests also performed for prediction of the mudrock minerals and chemical composition.

4 Results of Laboratory Tests

Hydrometer Tests for Mudrock Samples

Clay percentages of the different types of samples are performed to classify mudrock samples. Hydrometers tests are performed on dry fine materials passing sieve No.200. Classification was defined according to (Potter et al. 1980) method, which define that clay percentage is ranged between 24–36% for the tested mudrock samples. The clay percentage is important value for predicting the mudrock activity. Activity which is defined as the ratio of the plasticity Index (liquid limit (LL) minus plastic limit (PL)) to percent of clay fraction finer than 0.002 mm is one means of classifying expansive soils based on their index property. The activity index defines the capability of mudrock to swell. According to [Skemton 1953] clays are classified with respect to their activity, as

Table 1. Classification of expansive soil according to activity

Degree of activity	Activity
Inactive	less than 0.75
Normal	0.75–1.25
Active	greater than 1.25

Table 2. Index properties of Mudrock tested samples

Sample no.	Depth (m)	LL (%)	PL (%)	PI (%)	Free swell (%)	Activity
S1	4.0	48	20	28	110	1.4
S2	4.0	47	19	28	95	1.3
S3	4.5	53	19	34	125	1.6
S4	4.5	54	21	33	120	1.5

shown in Table 1. The activity of the tested samples shows that their activity index is more than 1.25 as shown in Table 2, which describe that the mudrock formations of the study area is in active state to swell.

Atterberg Limits Tests

Atterberg limit tests on mudrock samples are performed. The plastic limit (PL), Liquid limit (LL) and plasticity index (PI) of the tested samples is shown in Table 2.

Free Swelling test

The free swelling test is one of the most commonly used simple tests for estimating soil swelling potential. This test is performed by pouring 10 cm^3 of dry soil, passing through sieve no 40 (0.425 mm diameter), into a 100 cm^3 graduated cylinder. The cylinder is then filled with distilled water and the swelled volume of the soil is measured after the material settles [Egyptian code-part 2, 2001]. Where, free swelling is then given by:

$$Fs = (V - Vo)/V_O \times 100$$

Where:-

Fs = Free swelling, V = Final volume after swell and V_O = Volume of dry soil, 10 cm^3

Results of the free swelling tests are given in Table 2. The results illustrated that the free swelling ratio of the studied samples is ranging from 95 to 125%, and that means that the mudrock formation have the ability to swell as water reached to it in the future.

According to the change in volume of mudrock formation as water reached, there is an expectation of superstructures defects due to the clay activity in the mudrock is intermediate to high plasticity.

Mineralogical Composition

X-ray diffraction method was utilized for identification of the different types of clay minerals present in the tested mudrock samples. The mineralogical analysis of representative sample was carried out using PW1710 BASED diffractometer with a generator operating at 40 kV, 30 mA. The mineralogy of studied mudrock samples provides the basis for understanding their geotechnical behavior. It is also helps to identify types of their clay minerals such as high activity smectite minerals, calcite, quartz and other minerals. It was found that, the clay minerals present in the study area fraction were montmorillonite as major constituent, as well as kaolinite and discrete illite as minor constituents. The clay as montmorolonite in the soil is a good indication of the swell potential. These problems of excessive expansive characteristics lead to much damage to the structures built in and on these types of mudrock formation. The test results are shown in Fig. 4 for the different minerals.

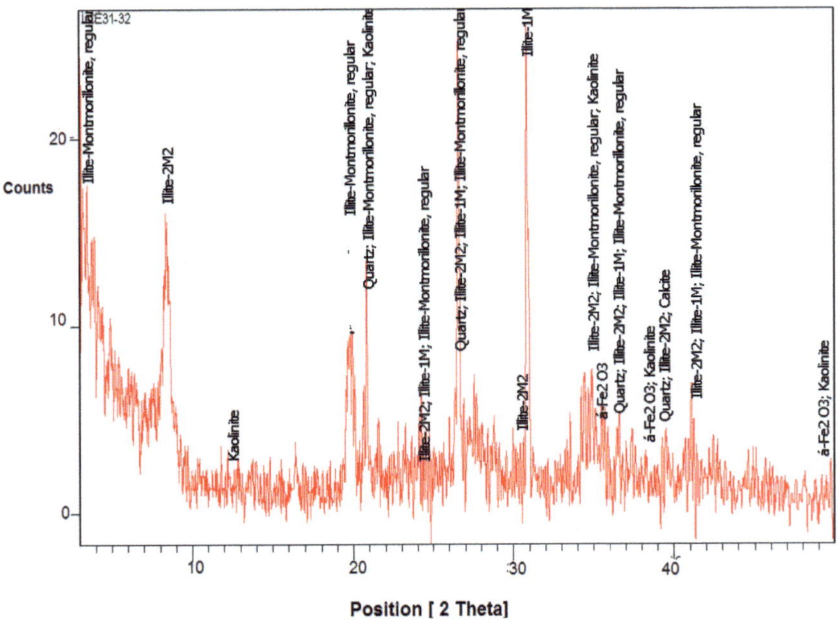

Fig. 4. X ray diffraction results for mudrock of Al Wadi Al Gadid

Chemical Analysis

Chemical analysis is carried out by X-ray fluorescence spectrometer (XRF) technique using the instrument JEOL, JSX 3222, Japan. Chemical analysis of studied samples indicated that they have high content of aluminum oxide (Al_2O_3) and ferric oxide (Fe_2O_3), which are the main elements of clay minerals and where the increasing of swelling strain of these samples related to the increasing of aluminum oxide (Al_2O3) and ferric oxide (Fe_2O_3), the percentage of chemical composition of (Al_2O3) and (Fe_2O_3) are 14.60% and 7.73%, respectively as shown in Table 3.

Swelling Pressure Tests

Swelling pressure (Sp) of a soil is the external pressure that needs to be placed over a swelling soil to prevent volume increase. The most reliable means of measuring

Table 3. Results of chemical analysis of the mudrock

Chemical composition											
SiO_2	Al_2O_3	Fe_2O_3	CaO	MgO	SO_3	Na_2O	K_2O	TiO_2	P_2O_5	Cl-	L.O.I
50.80	14.60	7.73	2.04	4.79	0.21	0.52	3.07	0.95	0.22	0.32	14.50

swelling pressure is laboratory determination using one-dimensional consolidometer. Three different test methods are used for the determination of swelling pressures to the site mudrock samples according to (Egyptian code- part 5, 2007). These methods are called the direct measurement for the swelling pressure.

The first method is by using the different Pressure and this method is the closest way to represent what happens in nature. Swelling pressure in this method is defined as the stress required keeping the sample size of the soil stable after being fully saturated with water under this stress. In this method, three or more identical samples are prepared and each sample is loaded with different stress. The samples are submerged and their vertical movement is monitored until the increase in volume ceases. The relationship between the different applied pressure with their heave values is drawn as shown in Fig. 5a. By extending the obtained curves to the zero heave value as shown in Fig. 5b, the swelling pressures of the tested mudrock samples will be obtained after drawing the applied pressure using the logarithmic scale. According to the different pressure method, the swelling pressure was ranging between 1020 kPa and 1250 kPa for the S3 and S1 mudrock samples, respectively as shown in Fig. 5-b.

The second method is performed by using the pre-swelling method by applying external pressure stresses that must be exerted on a swell mudrock sample until it returns to its original size. In this method, one sample is installed in the Oedometer device and submerged and allowed to swell under the influence of a small value pressure of 10 kPa.

The vertical movement is monitored until the increase in volume ceases. Then, external pressures are increased until the sample is compressed again and returned to its original size and this final pressure represents the swelling pressure.

Figure 6 illustrated that the swelling pressures of the three tested mudrock samples S2, S3 and S4, respectively. The swelling pressures using the pre-swelling method was ranged between 950 kPa, 1220 kPa and 1400 kPa for S2, S4 and S3, respectively.

While, the third test is performed by using the method of constant volume as the confined sample is immersed in water and no volume change is allowed. The resulting maximum pressure at which no more stress value on immersed sample is needed with time defined as the swelling pressure. This method is the least representative of what happens in nature. In this method, one sample is used to control the size to remain constant when immersed in water. The vertical movement is prevented using external stress until the sample compressed. Computerized Oedometer apparatus was used to maintain no volume change with time. Figure 7a illustrated a swelling pressure of 1145 kPa was achieved by using the computerized constant volume tested mudrock sample. While, Fig. 7b illustrated that the heave was nearly neglected during the constant volume change method.

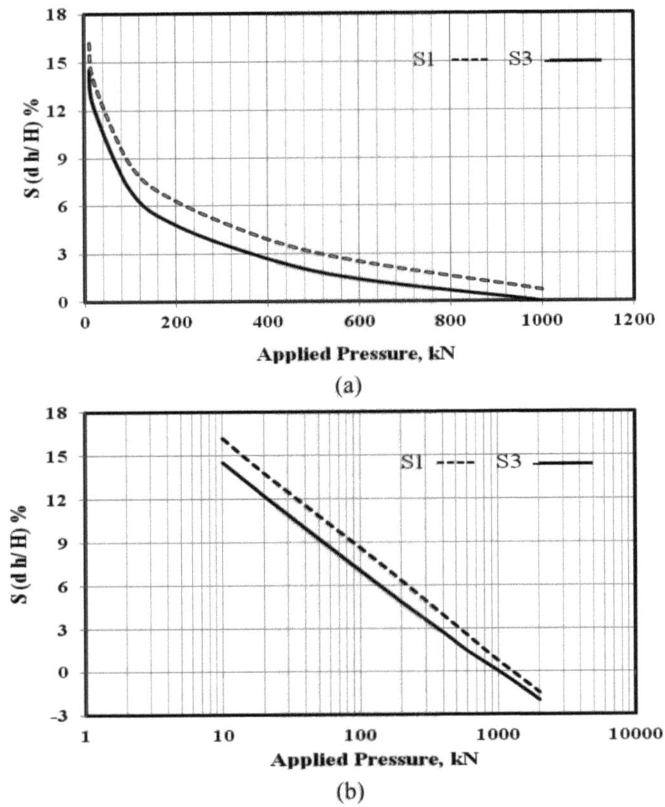

Fig. 5. (a) Relationship between applied pressure and swell potential using different pressure method for S1 and S3 mudrock samples (b) Swelling pressure of mudrock using the different pressure method for S1 and S3 mudrock samples

Swell Potential Testes

Swell potential (S %) is defined as the percentage increase in the original height of the specimen. In the swell potential test, the specimen is placed in the oedometer set under a small pressure of nearly about 10 kN/m^2, water is then added to the specimen and the expansion of the volume of the specimen is measured until an equilibrium is reached. The swell potential is calculated as follows:

$$S(\%) = (\Delta H)/H) \times 100$$

Where:

ΔH = Height of swell due to the saturation, H = Original height of the specimen.

From the swelling pressure tests of the different pressure and pre-swelling methods classified above, the value of the swell potential was determined while the third method of constant volume has no swell potential as it was maintain to be nearly zero during the test. Figures 5a and b illustrated that, the swell potential value using the different

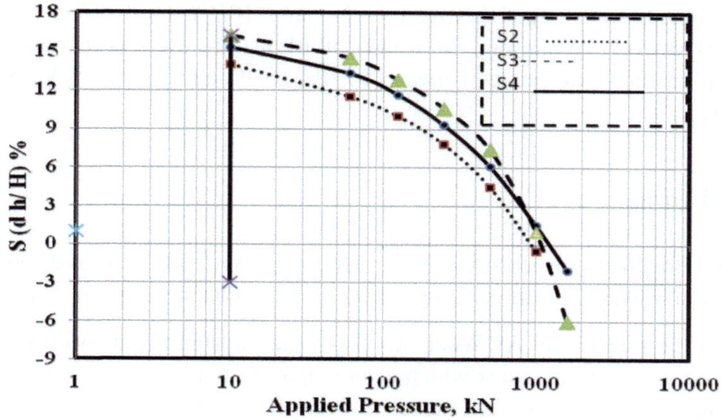

Fig. 6. Swelling pressure using pre-swelling pressure method for S2, S3, and S4 mudrock samples

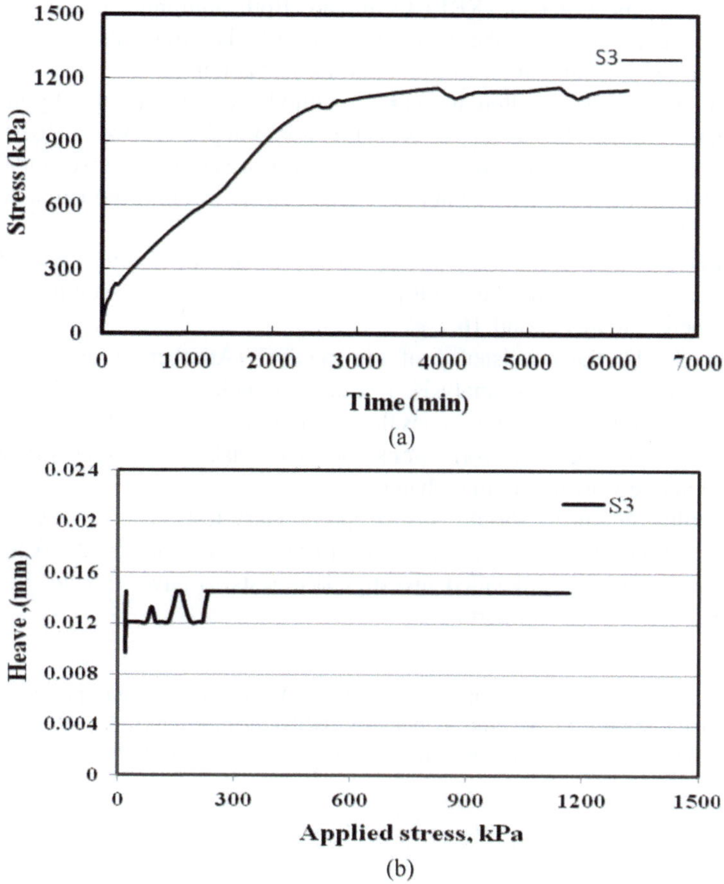

Fig. 7. (a) Swelling pressure of mudrock using constant volume method (b) Applied pressure on mudrock sample during the constant volume method

pressure method, was ranged between 14.5% and 16.3% for S1 and S3 mudrock samples, respectively. While Fig. 7a, b and c illustrated the swell potential value using pre-swelling method, was ranged between 14.1%, 15.3% and 16.2% for S2, S4 and S3, respectively.

5 Conclusions

In our study for the mudrock or shale formations of Al Wadi Al Gadid region at the central part of western desert of Egypt, the following conclusions are summarized:

1. Mudrock formation of Al Wadi Al Gadid has a significant swelling potential according to its activity index which is more than 1.25 as shown in Table 2 and also as its free swelling ratio was ranged between 95% and 125%.
2. Chemical analysis of mudrock formation indicated that it has a high content of aluminum oxide (Al_2O_3) and ferric oxide (Fe_2O_3) where their percentages was about 14.60% and 7.73%, respectively from the tested sample using X-ray fluorescence spectrometer (XRF) technique. Both aluminum oxide (Al_2O_3) and ferric oxide (Fe_2O_3) are the main elements of clay minerals which cause the increasing of swelling strain of the mudrock formation.
3. According to the three different pressure method performed on the dry undisturbed mudrock sample, the swelling pressure was ranging between 1020 kPa and 1250 kPa, while using the pre-swelling method it was ranged between 950 kPa and 1400 kPa. In the constant volume test method, the swelling potential was nearly about 1145 kPa.
4. The swell potential value using the different pressure method was ranged between 14.5% and 16.3%, while the swell potential value using pre-swelling method was ranged between 14.1% and 16.2%.
5. Combined with the consistency of the material (Atterberg limits/Activity), free swell tests and chemical analysis, it can be concluded that in all likelihood a Calcium Montmorillonite dominates the clay fraction.and expected to swell as water reached to the underneath formations and cause risk to any superstructures constructed above it as its volume change.
6. Geotechnical engineers should perform preliminary tests such as clay content and attereberg limits for mudrock formations of the study region of Al Wadi Al Gaid before performing any detailed investigations tests to have knowledge about the swell behavior of the formation.

Acknowledgments. The author appreciates Engineer Hayel Elnaggar and physical laboratory staff of the Geotechnical Engineering Laboratory (GEL) of the Housing and Building National Research Center (HBRC) for their help in performing the laboratory tests.

References

Awad, G.H., Ghobria, M.G.: Zonal stratigraphy of El-Kharga oasis. Geol. Surv. of Egypt, Paper No. 34 (1965)

Blatt, H., Middleton, G.V., Murray, R.C.: Origin of Sedimentary Rocks, pp. 782–789. Prentice Hall, Inc., Englewood Cliffs, New Jersey (1980)

Dhowian, A.W.: Characteristics of Al-Ghatt collapsing and swelling soils. In: Proceedings of Symposium on Geotechnical Problems in Saudi Arabia, vol. 1, pp. 3–33 (1981)

Dhowian, A.W.: Characteristics of expansive soils in the Northern Region of Saudi Arabia. In: Proceedings of the 5th International Conference on Expansive Soils. Adelaide, South Australia (1984)

Dick, J., Shakoor, A.: Litho logic controls of mudrock durability. Q. J. Eng. Geol. **25**, 31–46 (1992)

Egyptian code: Egyptian code of soil mechanics. In: Foundations Carrying Out and Designation, Part 2, Laboratory Tests, Egypt (2001a)

Egyptian code: Egyptian code of soil mechanics. In: Foundations Carrying Out and Designation, Part 5, Problimatic Soils, Egypt (2001b)

Fam, M., Dusseault, M., Fooks, J.: Drilling in mudrocks: rock behavior issues. J. Petrol. Sci. Eng. **38**, 155–166 (2003)

Grainger, P.: The classification of mudrocks for engineering purposes. Quat. J. Eng. Geol. **17**(4), 381–387 (1984)

Ghobrial, M.G.: The structural geology of the Kharga oasis. Geol. Surv. Egypt, paper 43 (1967)

Greene, B.: Predicting the unconfined compressive strength of mudrocks for design of structural foundations. Unpublished Ph.D., Dissertation, Department of Geology, Kent State University, Kent, Ohio, pp. 271 (2001)

Holtz, W.G., Gibbs, H.J.: Engineering properties of expansive clays. Trans. ASCE **121**, 641–663 (1956)

Issawi, B., Jux, U.: Contributions to the stratigraphy of the Paleozoic Rocks in Egypt. *Geology Surv.*, 28–39, Egypt (1982)

Johnson, L.D., Sherman, W.C, Mcanear, C.L.: Field test sections on expansive clays. In: Proceedings of 3rd International Conference on Expansive Soils, vol. 1, pp. 239–248. Haifa, Israel (1973)

Kostandi, A.B.: Eocene facies maps and tectonic interpretation in the western desert, Egypt. Rev. Inst. Franc. Pet.E. **18**, 1331–1343 (1963)

Mohamed, A.A.: Engineering characteristics of swelling clays. M.Sc. Thesis, Assiut University, Assiut, Egypt (1978)

Radwan, A.M., Abdel-Rahman, A.: Effect of macrofabric on the behavior of expansive clays. In: Proceedings Al-Azhar Engineering 1st Conference, vol. 4, pp. 111–120. Cairo (1989)

Said, R., Kerdany, M.T.: The geology and micropaleontology of Farafra oasis. Micropaleontology **7**, 317–336 (1961)

Sobolevsky, Y.A., Morgachiev, V.N.: Filtration anisotropy of swelling foundation soils and slopes. In: Proceedings of 6th International Conference on Expansive Soils, vol. 1, pp. 177–182. New Delhi, India (1973)

Potter, P.E., Maynard, J.B., Pryor, W.A.: Sedimentology of Shale, p. 306. Springer-Verlag, New York (1980)

Radwan, A.M., Abdel-Rahman, A.: Effect of macrofabric on the behaviour of expansive clays. In: Proceedings Al-Azhar Engineering 1st Conference, vol. 4, pp. 111–120. Cairo (1989)

Sarman, R., Shakoor, A., Palmer, D.: A multiple regression approach to predict swelling in mudrocks. Bull. Assoc. Eng. Geol. **31**(1), 107–121 (1994)

Skempton, A.W.: The Colloidial activity of clays. In: Proceeding 3rd International Conference on Soil Mechanics and Foundation Engineering, vol. 1, pp. 57–61. Switzerlandz (1953)

Sridharan, A., Prakash, K.: Classification procedures of expansive soils. In: Proceedings of Institution of civil Engineers, Geotechnical Engineering, vol. 143, pp. 235–240 (2000)

Case Study of Rupture and Recovery in Excerpt from BR-060 in the Municipality of Alexânia in the State of Goiás, Brazil

Rideci Farias[1(✉)], Tiago Martias Lino[2(✉)], Haroldo Paranhos[1(✉)], Itamar de Sousa Bezerra[3(✉)], Ranieri Araújo Faria Dias[4(✉)], Alexsandra Maiberg Hausser[5(✉)], and Roberto Pimentel de Sousa Júnior[6(✉)]

[1] Reforsolo Engineering, Brasília, Brazil
rideci.reforsolo@gmail.com, reforsolo@gmail.com
[2] Iesplan, Brasília, Brazil
tiagoltmatias@gmail.com
[3] Maccaferri, Brasília, Brazil
Itamar@maccaferri.com.br
[4] University of Pará, UFPA, Belém, Brazil
ranierileislie@yahoo.com.br
[5] UniCEUB, Brasília, Brazil
alexsandramaiberg@hotmail.com
[6] University of Brasília, UnB, Brasília, Brazil
eng.robertopimentel@gmail.com

Abstract. The BR-060/GO rupture occurred on the border between the Federal District and the state of *Goiás*, near the city of *Alexânia*/GO. The landslide occurred in the early hours of December 28, 2010 at km 24 of BR-060. On average, the highway passes about 60 thousand vehicles per day. The region is characterized by *cerrado* vegetation and rugged terrain with hills, valleys and lowlands, which contributed to this. The region is also surrounded by several water courses. The landslide resulted in a crater about 60 meters long and 12 meters deep, completely banning road and vehicular traffic in the region. Despite the seriousness of the accident, fortunately there were no casualties. The emergency works began on December 28, 2010. According to the National Department of Transportation Infrastructure (DNIT), these highway recovery works had a 180-day deadline. The contract also included the execution of the deviations for the liberation of vehicle traffic. The competent and supervising body of the work was the DNIT itself, in which it gave all the support and accompaniment to the execution of the services. In addition to the recovery of the slopes, these works had as objective the execution of the surface and underground drainage devices, as part of the solution. The underground drainage was characterized by the execution of deep drains, sub-horizontal drains and draining layers. Already the superficial drainage by the execution of gutters of central bed, gutters of sidewalk, descents of water, tubular manhole of concrete and boxes collector. For the recovery of the embankment, gabion containment, paving and environmental recovery services were also carried out. Therefore, this article aims to present analyzes of the possible causes of landslide, but also the solutions adopted to recover the damaged section. The main solution

J. Wasowski and T. Dijkstra (Eds.): GeoMEast 2018, SUCI, pp. 135–148, 2019.
https://doi.org/10.1007/978-3-030-02032-3_14

includes the gabion restraint system, associated with complementary works to recover the ruptured section, but also the consequent release of traffic on the road.

1 Introduction

This article aims to present an analysis of the possible causes of landslide, but also the solutions adopted for the recovery of the stretch that broke from BR-060/GO on the border between the Federal District and the state of Goiás, near the city of Alexânia/GO. The rupture occurred in the early hours of December 28, 2010 at km 24 of BR-060. On average on the highway it passes about 60 thousand vehicles per day. The main solution for recovery of the damaged section includes the gabion restraint system, associated with complementary works to recover the stretch for the release of traffic on the road.

2 Location of Damaged Stretch

The section that suffered rupture is located near the border of the Federal District and the state of Goiás, near the city of Alexânia/GO. The region is characterized by cerrado vegetation and rugged topography with hills, valleys and lowlands, and areas subject to runoff and water courses. Figure 1 shows the general location map of the area where the highway rupture occurred, which is approximately 15 km from the city of Alexânia/GO.

Fig. 1. Localization of BR 060/GO disruption (DNIT 2011).

The landslide resulted in a crater near 60 meters in length and 12 meters in depth, completely banning the route and traffic of vehicles in the region. The intense rainy season would be the main cause of the landslide, and fortunately no casualties, despite the seriousness of the accident. Figures 2, 3, 4, 5 and 6 show the section that broke.

Due to what happened there was a need to divert the traffic in two alternatives to the driver who left the Federal District to Goiânia. One by BR-070, towards Cocalzinho, to GO-414 and then to BR-060, which increases the route to the capital of Goiás by 28 km. The other option was to reach the GO-010 by Luziânia and continue until

Fig. 2. Broken section of BR-060/GO (12/31/2010).

Fig. 3. Broken section of the BR-060/GO (12/31/2010).

Fig. 4. Broken section of the BR-060/GO (12/31/2010).

Fig. 5. Broken section of the BR-060/GO (12/31/2010).

Fig. 6. Broken section of the BR-060/GO (12/30/2010).

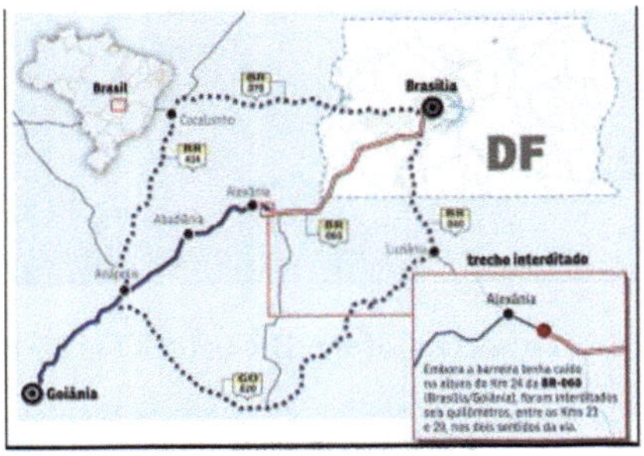

Fig. 7. Alternative paths when the BR-060 interdiction.

Goiânia. In this case, it is 50 km more, bringing the distance between Brasilia and Goiânia to 250 km. Figure 7 shows alternative routes for motorists when the BR-060 interdiction.

3 Analysis of Possible Causes

3.1 High Pluviometric Index - Soil Saturation

The rainy period in the region, with great intensity between the months of December and January, increased soil saturation, leading to erosive processes, mainly internal erosion, but also an increase in the value of percolation and the consequent decrease of soil resistance at sites where the flow of water was more concentrated. In addition, upstream of the cut made to install the highway, there is a considerable area in which the native vegetation was suppressed, aiming investments in agriculture and livestock. In addition to the suppression of vegetation, it was verified that the basin was placed in the ground for water storage, which, when reconciled with the infiltration and runoff, caused the possible preferential flow path in the soil mass to increase the saturation of the subsoil to the highway. Thus, it was pointed out as one of the main causes of the rupture of the road, the degree of saturation of the soil and the increase of the pore pressure, caused by the rains, with the consequent destabilization of the massif.

Figure 8 shows the possible flow inside the bulkhead, and Fig. 9 shows a view of the area with the infiltration basin upstream of the stretch of the highway that has undergone the rupture.

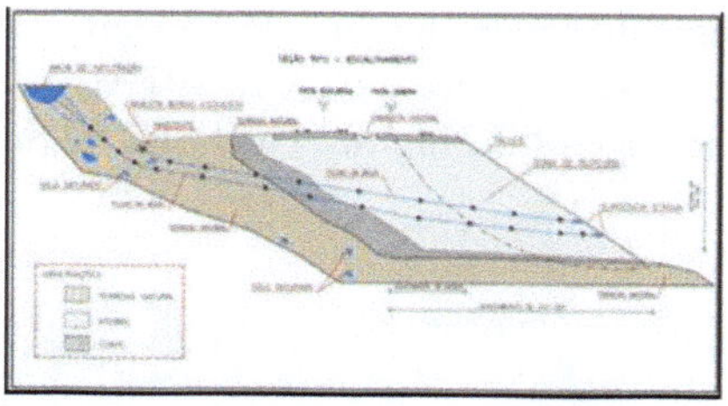

Fig. 8. Possible flow inside the massif (Lino 2016).

Fig. 9. Infiltration basin upstream of the highway (Lino 2016).

Fig. 10. View of saturated soil after landslide, December 2010 (Lino 2016).

Figure 10 shows the saturation inside the massif that with the excavations for soil removal, the flow of water was visible.

Despite some data and initial characteristics of the work, one could not say what had really caused the highway breakup. For this, more detailed investigations of the region were carried out, as follows.

3.2 Geotechnical Investigation

In addition to inspections in the area of the accident, it was also carried out for the execution of a SPT type survey in order to verify the conditions of the subsoil, groundwater of the region, besides assisting in the basic design studies.

At first there were three drill holes totaling 45.35 (forty-five meters and thirty-five centimeters). These probes were named SPT 01, SPT 02 and SPT 03, as shown in Fig. 11, with SPT 1 at the edge of the highway and SPT 02 and SPT 03 on the axis of each of the tracks.

Fig. 11. Location of SPT drill holes (BR-060, km 24).

Fig. 12. Execution view of the SPT 1 survey (BR-060, km 24).

Figure 12 shows the execution of the SPT 01 probe at the edge of the highway.

When considering the type of soil found, it was found to be predominantly a layer of landfill + material of the estradiol body with a variable thickness of up to three meters. Underlying this layer, clayey to slightly clayey/sandy to slightly sandy, varying up to 3.0 meters thick. Then, clay-silt/sandy clay, little clay. In terms of water level the SPT 01 probe was 1.50 meters from the “mouth” of the hole. In SPT 02 at 6.10 meters, and at SPT 03 at the depth of 7.15 meters. In view of this, it was possible to establish the possible behavior of the groundwater table in the massif.

4 Stability Analysis of Slope

For the proper analysis of slope stability, the on-the-spot checks were added to correlations with the SPT probes to choose the parameters.

In the simulation with the unsaturated soil, 10 kPa was used for cohesion, 30° for the friction angle and 17 kN/m^3 for the specific soil weight, obtained from correlations. Figure 13 shows the slope stability safety factor of 1.454. Simulation done with the empirical data of the dry soil, value according to the minimum criteria of safety factors of NBR 11682/2009.

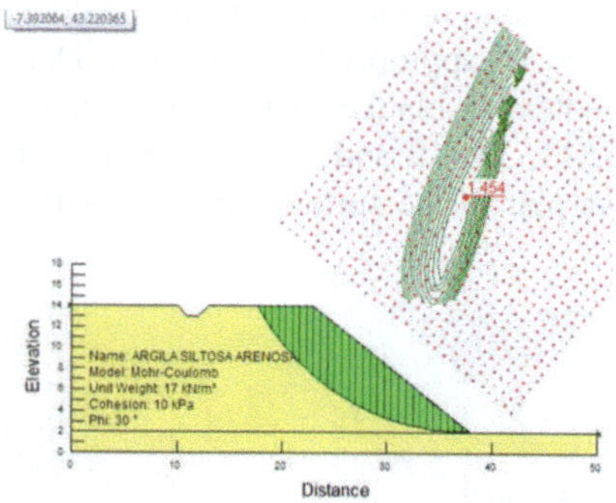

Fig. 13. Stability analysis with natural soil. (FS = 1.454).

In the second simulation of the slope stability, we have as a characteristic the calculation of the safety factor with the saturated soil. Also the correlated SPT data with empirical tables. For saturated soil, we worked with 0 (zero) kPa for cohesion, remaining 30° for the friction angle, although it was lower than the safety factor, and 17 kN/m^3 for the specific soil weight. Figure 14 shows the safety factor of 1.117. Note the expected reduction of the safety factor with the cohesion of 0 kPa, considering the saturated soil. Although the safety factor is greater than 1, it is not in accordance with the safety criteria of NBR 11682/2009.

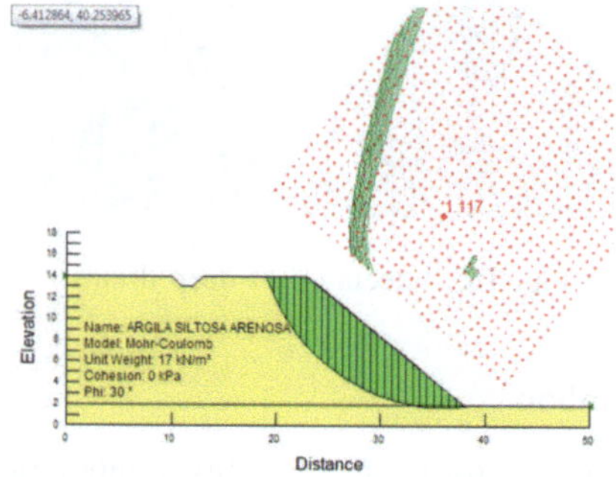

Fig. 14. Stability analysis with saturated soil. (FS = 1.117).

5 Solutions for Recovery of the Damaged Area

The project developed by DNIT for the emergency execution of the work involved a series of solutions with surface and underground drainage, containment with gabion, paving, but also the environmental recovery of the area affected by the works.

5.1 Surface and Underground Drainage

The internal or subterranean drainage aims to remove the percolation water from the massif (flow in the soil pores), reducing percolation flow. In the progress of excavation and removal of saturated soil, one of the concerns of the construction company was to channel the flow of water to an appropriate destination. Deep drainage systems were used (deep drain with geotextile blanket, sub-horizontal drain, DHP, and drainage mattress). Figures 15, 16, 17, 18, 19 and 20 show part of the execution of the deep drainage. Figure 21 shows part of the surface drainage with the finished work.

Fig. 15. Excavations for the installation of deep drains.

Fig. 16. Execution of deep drains.

5.2 Containment in Gabon

A widely used solution, whether for landfill stability, containment, surface protection of slopes, water courses and highways. They are elements of prismatic or cylindrical form, constituted by a metallic net and filled with stones of hand or rolled pebbles, destined to

Fig. 17. Execution of deep drains.

Fig. 18. Drilling for DHP installation.

Fig. 19. Execution of the draining mattress.

the solution of geotechnical, hydraulic problems and of erosion control. Gabions can be, (gabion type box, gabion type Reno mattresses or gabion in bags). Gabions are able to drain the waters from percolation of the terrain and to resist all kinds of efforts, and above all, to work under traction thus eliminating one of the main factors of instability. To recover the landfill and to solve drainage problems, gabion structures were used. Four gabion structures were executed, three box type and one Reno type. Figures 22 and 23 show the typical section of the highway and then in Figs. 24, 25, 26 and 27 show the execution and some of the systems in gabion, and in Figs. 28 and 29 the finished section.

Fig. 20. Drainage mattress construction.

Fig. 21. Smooth waterfall running along the banks of the stretch recovered on Highway BR 060.

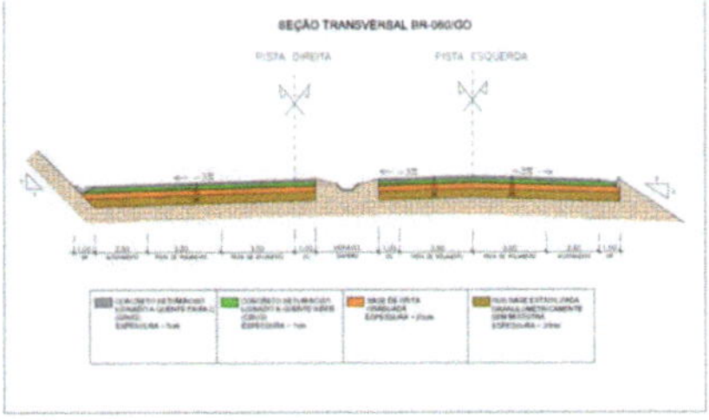

Fig. 22. BR 060 cross section in the ruptured section (DNIT 2011) (Geo-slope International 2012).

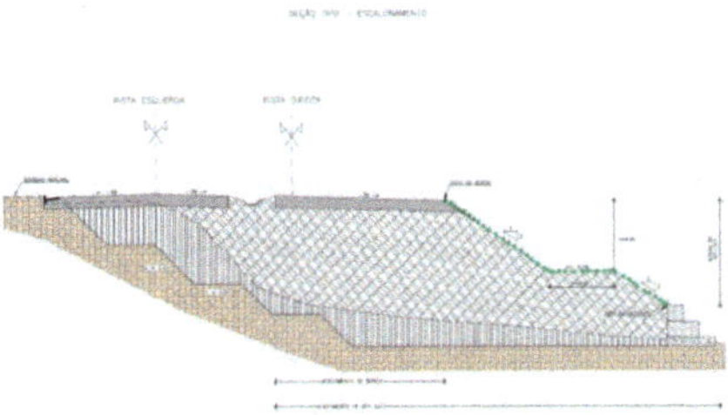

Fig. 23. Type section for the section retrieved from BR 060 (DNIT 2011) (Geo-slope International 2012).

Fig. 24. Execution of gabion restraint.

Fig. 25. Execution of gabion restraint.

Fig. 26. View of gabion restraint.

Fig. 27. System implanted with the completion of the work.

Fig. 28. Retrieved and completed section.

Fig. 29. Retrieved and completed section.

6 Conclusions/Recommendations

This article sought to present analyzes of the possible causes of landslide, but also the solutions adopted for the recovery of the stretch that broke from BR-060/GO on the border between the Federal District and the state of Goiás, near the city of Alexânia/GO. Thus, in the face of field studies and verifications, but also of due analysis, it is possible to conclude:

(a) The rainy season in the region, with great intensity between the months of December and January, increased the saturation of the soil, and the percolation with preferential paths of flow coming to cause erosive processes, mainly internal erosion, but also the consequent decrease of the resistance of the massif in places where the water flow regime was more concentrated;
(b) Thus, one of the main causes of road rupture was the degree of soil saturation and the increase in pore pressure caused by the rains, with the consequent destabilization of the massif;
(c) Upstream of the cut made for the installation of the highway, there was a considerable area in which the native vegetation was suppressed, aiming investments in agriculture and livestock. In addition to the suppression of vegetation, it was verified the installation of basin in the land for water storage that reconciled to the infiltration and to the runoff caused the possible preferential path of flow in the soil mass to increase the saturation of the subsoil to the highway;
(d) When analyzing possible causes of road rupture, the geological-geotechnical, environmental and hydrological studies of a region for the execution of engineering works are of great importance. With these studies one can determine the best form and the most appropriate type of work to be implemented;
(e) Interference and human action has caused negative aspects in the environment, consequently interfering in engineering works. These actions cause soil degradation, severe erosive processes and changes in the surface and underground dynamics of regions. In the case study in question, this interference caused the emergence of springs and infiltration basins, factors that contributed to the breakdown of the highway;
(f) With the analysis of the environmental problems found in the region, it is justified the execution of the drainage devices and the containment in gabions. Underground and surface drainage aims to conduct the water to suitable locations, avoiding damage to the implanted work. On the other hand, the gabion containment structures have as main objective to contain the landfill and avoid erosive processes;
(g) The importance of preventive measures instead of corrective ones in order to avoid damages in works and (or) disturbances to the population and damage to the environment;
(h) In order to verify the current conditions of the implanted system, the site where the rupture occurred was visited in July 2017 and it was verified that the implanted works are consistent with the expected operation.

Acknowledgment. The National Department of Transport Infrastructure (DNIT), JM Terraplenagem e Construções Ltda., Reforsolo Engenharia Ltda., Universidade Católica de Brasília (UCB), University Center of Brasília (UniCEUB) and the Institute of Higher Education Planalto with important contributions that made possible the accomplishment of this work.

References

ABNT NBR 11682:2009 - Estabilidade de encostas

Departamento Nacional de Infraestrutura de Transportes. DNIT: 2011. Projeto Executivo de Obra Emergencial para a Rodovia BR-060/GO, Div. DF/GO Entr° BR-364 (Contorno de Jataí), Entr° GO-425 – Entr° GO-139 (Início Travessia Urbana Alexânia)

Geo-slope International: Stress deformation modeling with SIGMA/W, an engineering methodology, 335 p. (2012)

Lino, T.M.: Estabilização de Taludes: Trabalho de Conclusão de Especialização, Publicação ENC. TCC– 020.1, Departamento de Engenharia Civil, Instituto de Ensino Superior Planalto, Brasília, DF, 82 p. (2016)

Use of the Pedological Map in the Geotechnical Characterization of Aris Mestre D'Armas in Planaltina - DF

Haroldo Paranhos[1(✉)], Rideci Farias[1(✉)], Joyce Maya Lucas Silva[2(✉)], Paulo Sergio Pereira[3(✉)], Roberto Pimentel de Sousa Júnior[4(✉)], and Alexsandra Maiberg Hausser[5(✉)]

[1] Reforsolo Engineering, Brasília, Brazil
reforsolo@gmail.com, rideci.reforsolo@gmail.com
[2] CONCREMAT, Brasília, Brazil
joyce.civil@gmail.com
[3] UCB, Brasília, Brazil
psergio@ucb.br
[4] University of Brasília, UnB, Brasília, Brazil
eng.robertopimentel@gmail.com
[5] UniCEUB, Brasília, Brazil
alexsandramaiberg@hotmail.com

Abstract. The present work presents the results of the geotechnical studies realized at ARIS Mestre D'Armas in Planaltina, DF, aiming to do a geotechnical characterization of soils and geotechnical risk, using field tests, SPT tests, ring penetrometer probes and sampling probes, permeability tests in drill holes and soil characterization tests in the laboratory. As an auxiliary tool in the spatialization and definition of the tested points of the "Geotechnical Field Investigations" program, through the prediction of local soil characteristics, the pedological and geological maps were used for the incursion of the field teams to the main groups of soils indicated in these maps. Parallel studies were executed in the pedogenetic slopes of the area for the evaluation of toposequence. The overlapping of the geological and pedological maps of the area defined the units with similar geotechnical characteristics. The methodology adopted consists of the evaluation of the characteristics of the soils of the area, based on the interpretation and descriptions contained in the pedological and geological maps; defining a map of unconsolidated material of geotechnical interest, of the area in question. The evaluation includes the superposition of the geological maps and the pedological map, based on methodologies, in which adaptations of geological and pedological classifications are made for geotechnical purposes. Characterization and geotechnical parameters were defined from field and laboratory tests on deformed and undisturbed samples, taken from strategically chosen points, in order to confer a certain representativity of the area under study. A correlation of the soils pedologically defined, with the geotechnical classifications were made by considering the properties, such as: granulometry, mineralogy, Atterberg limits, etc. and use of morphological data such as horizons thickness, macrofactory, color, etc., added to the physical properties of resistance and permeability determined in each profile. The verification of the

J. Wasowski and T. Dijkstra (Eds.): GeoMEast 2018, SUCI, pp. 149–160, 2019.
https://doi.org/10.1007/978-3-030-02032-3_15

geotechnical properties of each horizon of interest, through tests was associated to the pedological type, generating the geotechnical map of the area in question.

1 Introduction

This paper presents the results of the geotechnical studies executed at ARIS MESTRE D'ARMAS in Planaltina/DF, with the aim of characterizing the soils by means of field tests, SPT tests, soil tests, permeability tests in drill hole and works office. The studies consisted in the execution of a program of "Geotechnical Field Investigations", spatialized in the area of interest, according to the main groups of soils indicated in the pedological and geological maps.

2 Scope of Work

Initially the works were conducted in the office, with the evaluation of the existing documents of the area in question.

Subsequently, using the geological and pedological maps for the evaluation of toposequence, suggested and proposed by Soares et al. (1995), were carried out with the possession of pedological maps (Scale 1: 400,000) and geological characteristics of the region (scale 1: 250,000).

Based on the above mentioned studies, a field trip was carried out to for execute the SPT, penetrometric and drill holes for the evaluation of the masses.

In the following, brief considerations are made on the types of soils and also on their resistance and permeability.

3 Location of the Study Area

Figures 1 and 2 below, taken from Google Earth on July 10, 2013, show approximate macro views of the geotechnical survey and survey points, comprising 2625.00 ha.

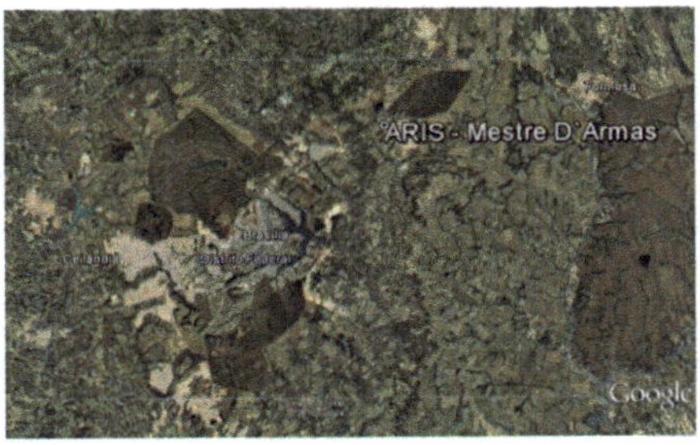

Fig. 1. Macro view of the area of execution of geotechnical studies in Planaltina - DF.

Fig. 2. Macro view of the area and location of the points studied.

4 The Use of Pedology in Geotechnics

Several other interpretations can be made from soil surveys, in addition to agricultural purposes, Larach (1983) cited by Resende et al. (1995). Baize et al. (1990) affirm that with the creation of a pedological referential, specialists from other scientific areas will be able to construct another typology, corresponding to their own needs, result of the horizons of reference.

4.1 Applied Pedology

Many works are cited by Paranhos (1998) correlating several attributes of pedology with geotechnical engineering. The overlap of geological and pedological maps of a given area can define units with geotechnical characteristics.

4.2 Pedogenesis

The pedology, based on models, indicates five main soil-forming factors and their properties are caused by climate, organisms, source material, time and topogaphy.

4.3 Soil × Landscape

The landscape is the image of the combined action of soil formation factors, such as topography, organisms, origin material, climate, over time. It is very important to know the distribution of the soils in the landscape in the execution of soil (or pedological) surveys, and also in studies of soil genesis (Prado 2005).

4.4 Local Soils

Figure 3 shows the soil group of the region under study and Fig. 4 shows the geological formation of the region.

Generally, at the top of the geopedological slope, there is the presence of more mature soils (Dark Red Latosols and Yellow Reds), which have a high drainage

LEGENDA

LV – Red latossoil;

LA – Red Yellow Latosoil

CX – Cambissoil;

GX – Gleissoil.

Fig. 3. Soil groups of the region (Modified - EMBRAPA, 2004).

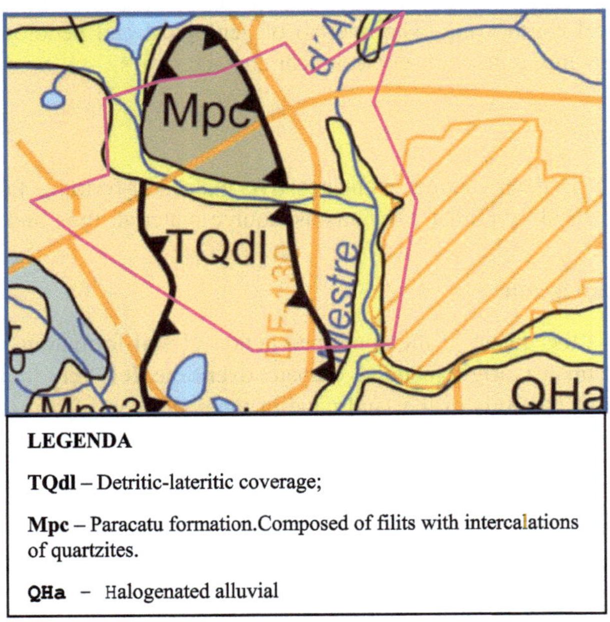

LEGENDA

TQdl – Detritic-lateritic coverage;

Mpc – Paracatu formation.Composed of filits with intercalations of quartzites.

QHa – Halogenated alluvial

Fig. 4. Geological Formation of the region (Modified - EEZ (2002).

capacity. The cambisols arise in specific conditions of the slope, associated to the inclinations and the lack of the action of the weathering.

It is also observed the presence of soils that had in their formation process the conditions of excess water and low hydraulic gradient (hydromorphism). There are also dark plastic soils with a high content of organic matter, typical of regions of accumulation of vegetal sediments.

5 Methodology

The methodology consists of the evaluation of the soil characteristics, from the interpretation and superposition of the geological and pedological maps. This methodology is based on works described and presented by Dias and Milititsky (1990), Bastos (1991) cited by Dias and Milititsky (1994) and Paranhos (1998), in which adaptations of geological and pedological classifications for geotechnical purposes are made.

In general, the technical bulletins that accompany the pedological maps provide important information for the different soil classes, taking into account the following criteria: B horizon type, base saturation percentage, clay activity, type of soil horizon A, textural class and drainage class, besides the phase criterion, taking into account the vegetation, topography, stoniness and presence of lateritic concretions. This bulletin also contains a detailed description of the main morphological characteristics (color, texture, structure, porosity, etc.).

Characterization and geotechnical parameters can be defined from field and laboratory tests in samples representative of the area under study.

Aerial photographs, geological maps, topographic maps and other related works can be used to better detail the areas. This type of investigation, carried out in an "office", serves to avoid overlapping fieldwork, making the visits exhaustive; (geological and geological mapping), which is the fundamental idea of the proposal that involves the present work: an interpretation of the behavior of the soil in an indirect way, based on already existing works (geological and pedological maps). The estimate of geotechnical units will assume initially the symbology of work published by Camargo et al. (1987), also cited by Dias (1995), to classify the surface horizons and to the subsurface classification, that is, horizons that keep the geological structure, will be adopted the simplified geological classification, based on the work of IAEG (1979).

The following symbols for the geotechnical classification were adopted:

$$ABxyz \tag{i}$$

Which:

- The capital letters for superficial horizons.
- Lowercase letters for subsurface horizons.

6 Obtaining Geotechnical Characteristics of Aris Mestre D'Armas Surface Soils, in Planaltina-Df, Using Pedological Survey

The geotechnical characteristics of the soils under study (Fig. 5) were estimated using the overlapping of the EMBRAPA (2004) pedological maps, in a scale of 1: 400,000 and geological ZEE (2002) in a scale of 1: 250,000, to facilitate fieldwork; rationally defining sample collection points and polls in general.

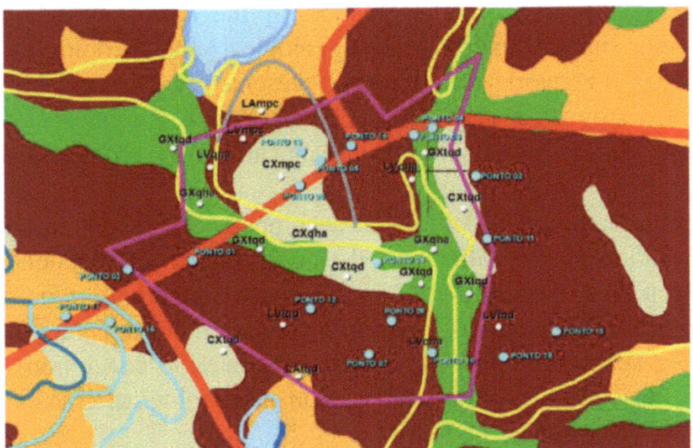

LEGENDA

LV_{TQd} – Red Latossoil, Detritic-lateritic coverage;

LV_{Mpc} – Red Yellow latossoil, Paracatu formation;

LV_{QHa} – Red Yellow Latossoil, Halogenated alluvial;

LA_{TQd} – Red Yellow Latossoil, Detritic-lateritic coverage;

LA_{Mpc} - Red Yellow Latossoil, Paracatu formation;

CX_{TQd} – Cambissoil, Detritic-lateritic coverage;

CX_{Mpc} - Cambissoil, Paracatu formation;

CX_{QHa} - Cambissoil, Halogenated alluvial;

GX_{TQd} – Gleissoil, Detritic-lateritic coverage;

GX_{QHa} - Gleissoil, Halogenated alluvial.

Fig. 5. Geotechnical Units of the region and location of the points studied.

6.1 Surveys and Field Tests

The field tests were performed in order to determine the soil resistance and permeability characteristics. The following tests were used: 14 probes, 14 probes with the ring penetrometer and 14 permeability tests.

6.2 Evaluation and Description of Study Areas

By means of the adopted methodology, field and laboratory tests, the geotechnical map in question (Fig. 5) and the appropriate descriptions of the geotechnical units are suggested.

Table 1 shows the soil variation found in the region by means of the overlapping of the pedological and geological maps, in order to generalize the geotechnical units.

Table 1. Pedological, geological and geotechnical units.

Pedological unit	Geological unit	Geotechnical unit
LV	TQdl	$\mathbf{LV}_{TQd}$
LV	Mpc	$\mathbf{LV}_{Mpc}$
LV	QHa	LV_{QHa}
LA	TQdl	$\mathbf{LA}_{TQd}$
LA	Mpc	$\mathbf{LA}_{Mpc}$
CX	TQdl	$\mathbf{CX}_{TQd}$
CX	Mpc	$\mathbf{CX}_{Mpc}$
CX	QHa	$\mathbf{CX}_{QHa}$
GX	TQdl	$\mathbf{GX}_{TQd}$
GX	QHa	$\mathbf{GX}_{QHa}$

6.3 Description of the Soils

The following is a description of the proposed geotechnical units.

LVTQd - Red Latosol, Detritus-Lateritic Cover
They are very deep soils, where the thickness of horizons A and B together, is greater than 3 meters, clayey or medium texture and rich in sesquioxides. They are very porous, quite permeable and markedly drained.

The susceptibility to erosion varies from practically null to light.

LVMpc - Red Latosol, Paracatu Formation
Idem to the LVTQd, but presenting smaller depths in the B horizon, due to the presence of the rock in subsurface. It is a transition zone between the Oxisols and the cambisols.

LVQHa - Red Latosol, Holocene Alluvial
Idem to the LVTQd but presenting small depths in the B horizon. It is a transition zone between the Latosols in the lowland regions.

LATQd - Red-Yellow Latosol, Detritus-Lateritic Cover
They are soils with high permeability of water. Very Deep Chemically, more than 95% of the lattosoils are dystrophic and acidic, with low to medium cation exchange capacity and pH levels around 4.0 and 5.5.

LAMpc - Red-Yellow Latosol, Paracatu Formation
Idem to the LATQd but presenting smaller depths in the horizon B, by the presence of the rock in subsurface. It is a transition zone between the Oxisols and the cambisols.

CXMpc - Cambisol, formation Paracatu
They are soils that have subsurface horizon with little physical and chemical alteration. It presents low degree of weathering. Generally, they are associated with more mountainous topographies (wavy and strong-wavy). They range from shallow and deep, reaching between 0.2 and 1 m.

CXTQd - Cambissolo, Detrito-Laterítica Cover
Idem to the CXMpc, but presenting some depths in the horizon B, by the presence of soil on the subsuperficial rock. It is a transition zone between the Oxisols and the cambisols in lowland regions.

CXQHa - Cambisol, Holocene Floods
Idem to the CXMpc, but presenting some depths in the horizon B, by the presence of soil on the rock in subsurface. It is a transition zone between the Cambisols and the Gleissolos in lowland regions.

GXQHa - Gleissoil, Holocene Mudslides
They are hydromorphic soils that generally occupy the depressions of the landscape, subject to floods. They present drainage of the types: poorly drained or very poorly drained, occurring frequently, thick dark layer of organic matter poorly decomposed on a grayish (gleaned) layer, resulting from oxidation environment.

GXTQd - Gleissolo, Detrito-Laterítica Cover
Same as GXQHa, but presenting some soil properties in the unsaturated zone. Capillary height generated by the presence of NA, does not reach this superficial layer. It is a transition zone between the Oxisoils and the Gleissoils in lowland regions.

6.4 Summary of Geotechnical Characteristics

Table 2 shows the classification of the soils of the geotechnical units.
Table 3 presents the classification for resistance.
Table 4 presents the permeability classification.

6.5 Susceptibility to Erosion

According to Camapum de Carvalho et al. (1993), the occurrence of erosions in the Federal District is related to two main erosive agents: man and water action, which generally act associated, although in different proportions and conditioned by soil erodibility.

Erosive Susceptibility in Studied Soils

Table 2. Soil classification of geotechnical units (SUCS, TBR and MCT).

Tipo de solos	Prof. (m)	SUCS	TBR	MCT
LV_{TQd}	1,0	CL e MH	A-7-5	LA'-LG'
	3,0	CL e MH	A-7-5	LA'-LG'
	5,0	MH	A-7-5 e A-7-6	LG'
	7,0	MH	A-7-5	LG'
LV_{Mpc}	1,0	CL e MH	A-7-5	LA'-LG'
	3,0	ML	A-4	NS'
	5,0	ML	A-4	NS'
LV_{QHa}	1,0	CL e MH	A-7-5	LA'-LG'
	3,0	SM	A-2-4	NA'
	5,0	SM-SC	A-3	NA
LA_{TQd}	1,0	CL e MH	A-7-5	LA'-LG'
	3,0	CL e MH	A-7-5	LA'-LG'
	5,0	MH	A-7-5 e A-7-6	LG'
	7,0	MH	A-7-5	LG'
LA_{Mpc}	1,0	CL e MH	A-7-5	LA'-LG'
	3,0	ML	A-4	NS'
	5,0	ML	A-4	NS'
CX_{Mpc}	1,0	ML	A-4	NS'
	3,0	ML	A-4	NS'
CX_{TQd}	1,0	CL e MH	A-7-5	LA'-LG'
	3,0	ML	A-4	NS'
	5,0	ML	A-4	NS'
CX_{QHa}	1,0	SM-SC	A-2-4	NA'
	3,0	ML	A-2-4	NA'
	5,0	ML	A-3	NA
GX_{TQd}	1,0	CL e MH	A-7-5	LA'-LG'
	3,0	SM-SC	A-2-4	NA'
GX_{QHa}	1,0	ML	A-4	NA'
	3,0	SM-SC	A-2-4	NA'

6.6 Red Latosols

The high velocity of water infiltration in these soils, avoids the formation of large volumes of surface runoff and associated with the low slope of the area, provides less susceptibility to erosion.

6.7 Yellow Red Latosols

This soil class is commonly found in water dividers and in transition areas for plateau ridges with flat to smooth corrugated topography, associated with a secondary field of soil with good drainage and permeability, not observing the development of processes erosive.

Table 3. Soil classification of the Geotechnical Units (SPT and Admissible Voltage).

Tipo de solos	Prof. (m)	SPT	Tensao admisslvel (kgf/cm^2)	Designafao
LV_{TQd}	1,0	2 a 5	0,4 a 1,2	Mole
	3,0	2 a 5	0,4 a 1,2	Mole
	5,0	6 a 10	1,4 a 2,2	Media
	7,0	6 a 10	1,4 a 2,2	Media
LV_{mpc}	1,0	2 a 5	0,4 a 1,2	Mole
	3,0	6 a 10	1,4 a 2,2	Media
	5,0	6 a 10	1,4 a 2,2	Media
LV_{QHa}	1,0	2 a 5	0,4 a 1,2	Mole
	3,0	6 a 10	1,4 a 2,2	Media
	5,0	6 a 10	1,4 a 2,2	Media
LA_{TQd}	1,0	2 a 5	0,4 a 1,2	Mole
	3,0	2 a 5	0,4 a 1,2	Mole
	5,0	6 a 10	1,4 a 2,2	Media
	7,0	6 a 10	1,4 a 2,2	Media
LA_{Mpc}	1,0	2 a 5	0,4 a 1,2	Mole
	3,0	6 a 10	1,4 a 2,2	Media
	5,0	6 a 10	1,4 a 2,2	Media
CX_{Mpc}	1,0	5 a 8	1,2 a 1,8	Pouco Compacta
	3,0	9 a 18	2,0 a 3,2	Median. compacta
CX_{TQd}	1,0	2 a 5	0,4 a 1,2	Mole
	3,0	5 a 8	1,2 a 1,8	Pouco Compacta
	5,0	9 a 18	2,0 a 3,2	Median. compacta
CX_{QHa}	1,0	5 a 8	1,2 a 1,8	Pouco Compacta
	3,0	5 a 8	1,2 a 1,8	Pouco Compacta
	5,0	5 a 8	1,2 a 1,8	Pouco Compacta
GX_{TQd}	1,0	0 a 2	< 0,4	Muito Mole
	3,0	2 a 5	0,4 a 1,2	Mole
GX_{QHa}	1,0	0 a 2	< 0,4	Muito Mole
	3,0	5 a 8	1,2 a 1,8	Pouco Compacta

6.8 Switching

These soils occur mainly on slopes and slopes with higher slopes. The high silt content of the A horizon and the restricted depth of the profile make this kind of soil permeability difficult. The combination of these features with the high slope rates where these soils occur makes them more susceptible to erosion.

6.9 Gleissoil

The limitations and the environmental sensitivity of these soils are very large, so much that they are considered as an area of environmental preservation. The limitations for

Table 4. Classification of soils of geotechnical units regarding permeability.

Tipo de solos	Prof. (m)	Ordem de permeabilidade	Designação
LV_{TQd}	1,0	10^{-4} a 10^{-2}	Permeável
	2,0	10^{-4} a 10^{-2}	Permeável
LV_{Mpc}	1,0	10^{-4} a 10^{-2}	Permeável
	2,0	10^{-4} a 10^{-2}	Permeável
LV_{QHa}	1,0	10^{-4} a 10^{-2}	Permeável
	2,0	10^{-4} a 10^{-2}	Permeável
LA_{TQd}	1,0	10^{-4} a 10^{-2}	Permeável
	2,0	10^{-4} a 10^{-2}	Permeável
LAMpc	1,0	10^{-4} a 10^{-2}	Permeável
	2,0	10^{-6} a 10^{-4}	Median. Permeável
CXMpc	1,0	10^{-6} a 10^{-4}	Median. Permeável
CXTQd	1,0	10^{-4} a 10^{-2}	Permeável
	2,0	10^{-6} a 10^{-4}	Median. Permeável
CXQHa	1,0	10^{-4} a 10^{-2}	Permeável
	2,0	10^{-6} a 10^{-4}	Median. Permeável
GXTQd	1,0	10^{-4} a 10^{-2}	Permeável
	2,0	10-8 a 10^{-6}	Pouco Permeável
GXQHa	1,0	10^{-4} a 10^{-2}	Permeável
	2,0	10-8 a 10^{-6}	Pouco Permeável

Table 5 Classification of soils of geotechnical units regarding erosive susceptibility.

Tipo de solos	Classe de Erodibilidade
LV_{TQd}	Muito Baixa a baixa
LV_{Mpc}	Muito Baixa a baixa
LV_{QHa}	Muito Baixa a baixa
LA_{TQd}	Muito Baixa a baixa
LA_{Mpc}	Muito Baixa a baixa
CX_{Mpc}	Forte
CX_{TQd}	Forte
CX_{QHa}	Forte
GX_{TQd}	Muito Baixa a baixa
GX_{QHa}	Muito Baixa a baixa

urban use are due to its high plasticity, common presence of organic matter and low permeability. Table 5 presents the summary of the erodibility class of the soils of the area.

6.10 Susceptibility to Flooding

For the area in question, the risk of flooding is only likely to occur in the regions delimited by soils classified as Gleissoils. These soils demarcate the strip with direct influence of streams and adjacencies.

6.11 Susceptibility to Colapse

The evaluation of soil collapse was done indirectly through the interpretation of SPT surveys in the area and the soil profile. In general, the collapsible soils are soils with the soft macroporous structure, with high porosity or high void index. They are usually unsaturated soils and have a low SPT in the assessment altitudes.

7 Conclusions

The generation of an unconsolidated material map, based on the overlapping of Pedological and Geological maps, for characterization of areas is a tool of great geotechnical potential, although the limit positions of the units and the transitions between them should be better evaluated by field and located probes.

The limitations of the present study are also highlighted, since databases were used at regional scales (1: 400,000 and geological scale on the 1: 250,000 scale).

References

Baize, D., Girard, M., Boulaine, J., Cheverry, C.L, Ruellan, A.: Um Referencial Pedológico. Porquê? (1990)

Dias, R.D., Militissky, J.: Metodologia de Classificação de Unidades e Perfis Geotécnicos Desenvolvida na UFRGS. SOLOS E ROCHA **17**, 81–92 (1994)

Dias, R.D.: Proposta de Metodologia de Definição de Carta Geotécnica Básica em Regiões Tropicais e Subtropicais, Revista do instituto geológico, volume especial, pp. 51–55 (1995)

Embrapa Cerados: Mapa pedológico digital – SIG atualizado do Distrito Federal escala 1.100.00 e uma síntese do texto explicativo, 31 p. (2004)

IAEG: Association for Engineering Geology and the Environment (1979)

Paranhos, H.S.: Caracterização Geotécnica dos Principais Grupos de Solos do Mapa Pedológico do Distrito Federal (Esc. 1:100.000): Estudos na Área de Dinamização Urbana, entre Samambaia e Gama (1998)

Prado, H.: Pedologia Fácil - Aplicações em solos tropicais (2005)

Resende, M., Curi, N., Rezende, S.B., Corrêa, G.F.: Pedologia: base para distinção de ambientes, Jard, Neput, Viçosa, 304 p. (1995)

Soares, A.M.L., et al.: Áreas degradadas suscetíveis aos processos de desertificação no Ceará. IPEA, Brasília (1995)

ZEE: Projeto Zoneamento Ecológico-Econômico da região integrada de desenvolvimento do Distrito Federal e Entorno – Mapa Geológico – Fase I (2002)

Author Index

J. Wasowski and T. Dijkstra (Eds.): GeoMEast 2018, SUCI, pp. 161–162, 2019.
https://doi.org/10.1007/978-3-030-02032-3

MIX
Papier aus verantwortungsvollen Quellen
Paper from responsible sources
FSC® C105338

If you have any concerns about our products,
you can contact us on
ProductSafety@springernature.com

In case Publisher is established outside the EU,
the EU authorized representative is:
Springer Nature Customer Service Center GmbH
Europaplatz 3, 69115 Heidelberg, Germany

Printed by Libri Plureos GmbH
in Hamburg, Germany